Presenting Science Concisely

Presenting Science Concisely

Dr. Bruce Kirchoff

Department of Biology, University of North Carolina at Greensboro, North Carolina, USA

Jon Wagner, Illustrator

Biology doodler extraordinaire, Portland, Oregon, USA

PUBLISHING

Originally published by CAB International, Wallingford, UK www.cabi.org
© Bruce Kirchoff 2021 (text), © Jon Wagner 2021 (illustrations)

A catalogue record for this book is available from the National Library of Australia.

ISBN: 9781486314683 (pbk)

This edition published exclusively in print only, in Australia and New Zealand, by:

CSIRO Publishing
Private Bag 10
Clayton South VIC 3169
Australia
Telephone: +61 3 9545 8400
Email: publishing.sales@csiro.au
Website: www.publish.csiro.au
Sign up to our email alerts: publish.csiro.au/earlyalert

Edited by Priscilla Sharland
Cover design by Jon Wagner, Illustrator and Sarah Hillier
Typeset by SPi, Pondicherry, India
Index by Archana Vinu
Printed in Australia by Ligare

CSIRO Publishing publishes and distributes scientific, technical and health science
books, magazines and journals from Australia to a worldwide audience and conducts
these activities autonomously from the research activities of the Commonwealth
Scientific and Industrial Research Organisation (CSIRO). The views expressed in this
publication are those of the author(s) and do not necessarily represent those of, and
should not be attributed to, the publisher or CSIRO. The copyright owner shall not
be liable for technical or other errors or omissions contained herein. The reader/user
accepts all risks and responsibility for losses, damages, costs and other consequences
resulting directly or indirectly from using this information.

CSIRO acknowledges the Traditional Owners of the lands that we live and work
on across Australia and pays its respect to Elders past and present. CSIRO
recognises that Aboriginal and Torres Strait Islander peoples have made and
will continue to make extraordinary contributions to all aspects of Australian
life including culture, economy and science.

The paper this book is printed on is in accordance with the standards of the Forest
Stewardship Council®. The FSC® promotes environmentally responsible, socially
beneficial and economically viable management of the world's forests.

Oct21_01

Contents

Introduction

Stories are the universal language. They are the way we create meaning. Read any scientific abstract and you will find a story. But what is a story? Even more importantly, what is a scientific story? This book will answer these questions in the context of short verbal presentations.

Chapter 1 frames the question and points out the similarity between story structure and the scientific process. These themes are explored more deeply in Chapter 2 where we look at three- and five-act structures and learn how these forms can be used to present scientific research. We also introduce Randy Olson's And-But-Therefore (ABT) format (Olson, 2020), which will play an important role in Chapters 4 and 5. Chapter 2 ends with a brief survey of short presentation forms, most generically known as lightning talks. Then in Chapter 3 we explore one of these forms, the Three Minute Thesis (3MT®) in detail (The University of Queensland (2008–present). We also talk about the importance of telling simple, concrete, emotional, and credible stories (Heath and Heath, 2007). We return to the And-But-Therefore format in Chapter 4 where we discuss extremely short presentations of 90 seconds or less. These elevator pitches are playing an increasingly important role in scientific communication. Chapter 5 rounds out our work with verbal presentations with a brief consideration of longer forms. We combine ideas from the five-act and And-But-Therefore forms to create presentations of up to an hour. In Chapters 6 and 7 we turn to poster presentations. Chapter 6 begins with a consideration of titles. We provide guidance on how to write a good title and how to use that title to focus the content of your poster. We also present and discuss good poster designs, including Mike Morrison's designs for a Better Scientific Poster (Morrison, 2021). In Chapter 7 we analyze some real posters from scientific meetings, including an analysis of eye-tracking data provided by EyeQuant (EyeQuant, 2009–present). In Chapters 8 and 9 we turn to a consideration of your audience and the presentation skills you need to make an impact. You are always addressing a specific audience, in a specific venue, on a specific topic. Failure to take this into account is one of the problems with many scientific presentations.

We have included exercises for the reader throughout the book. These range from putting research into three- and five-act structures in Chapters 2 and 5, to writing titles in Chapter 6, to perfecting your presentation skills in Chapter 9. You will get more out of the book if you do the exercises and apply them to your own presentations.

The author is available for workshops and consultations if you would like to work in more depth. Please contact him at kirchoffbruce@gmail.com.

Dr. Bruce Kirchoff

EyeQuant (2009–present) EyeQuant. Available at: https://www.eyequant.com/
Heath, C. and Heath, D. (2007) *Made to Stick*. Random House, New York.
Morrison, M. (2021) Better Scientific Poster. Open Science Framework (OSF), Center for Open Science, Charlottesville, Virginia. Available at: https://osf.io/ef53g/
Olson, R. (2020) *The Narrative Gym*. Prairie Starfish Press, Los Angeles, California.
The University of Queensland (2008–present) 3MT® Three Minute Thesis. The University of Queensland, Brisbane, Queensland, Australia. Available at: https://threeminutethesis.uq.edu.au/

Acknowledgments

This book would never have been written if it had not been for Kim Cuny and Rebecca Stubbs. Kim is the Director of the University Speaking Center at the University of North Carolina at Greensboro and has been wonderfully supportive of every aspect of scientific communication, and of this project. In addition to providing moral support, she wrote the first draft of Chapter 9 and contributed important ideas to other chapters. Rebecca is the acquisitions editor at CABI who approached me about writing a book about short scientific presentations. She has been wonderful to work with even when dealing with licensing issues, and that is saying something!

I also want to thank the scientists who consulted about their research and gave me permission to use their work. In addition to the citations in the chapters, Mike Morrison, Dr. Pieter Visscher and Professor David Nash require special mention as does Brian Palermo. They all went out of their way to provide helpful comments on the manuscript. I am especially grateful to Brian Palermo and Mike Morrison for reading and commenting on early drafts of several chapters. I also thank Mike for alerting me to Milan Klöwer's excellent poster, which I review in Chapter 7. I am also grateful to Alice Henebury, the Head of Marketing at EyeQuant, for allowing me to use their eye-tracking simulations in Chapter 7.

While writing the chapter on longer presentations I had the opportunity to attend Randy Olson's *ABT (And-But-Therefore) Framework "Course"* (http://storycirclestraining.com/). Participation in the course clarified my understanding of this form and helped me use it more effectively in Chapters 4 and 5.

Writing is time-consuming and can be draining. This book may never have been completed without the support and love of my wife, Mary Kirchoff. She put up with my poor moods, my complaints about lack of progress, and my poor sleep. She encouraged me to exercise, got me skiing, and taught me Pickleball. I am indebted to her on many fronts.

Despite the debt that I owe to the people who have supported this work, final responsibility for the book remains mine, as does responsibility for the errors.

First Principles: Explaining Science Through Stories

1

The greatest scientists are always artists as well.
(Albert Einstein cited in Henderson, 1955)

This book is about science and the art of storytelling. It is a book that builds on the commonality between the scientific process and narrative structure. It will help you become a better communicator. It will help you to embrace your artistic side and tell good stories about your work.

Although stories can be fanciful constructions, they can also relate true events with the artist's craft. Creative non-fiction is not fanciful. It adheres to the reality of events as closely as a scientist interpreting her results. This book is creative non-fiction. It uses factually accurate prose to describe scientific research and uses the storyteller's sensibility to present these facts. The author, a scientist, is also a storyteller and a teacher of storytellers. He brings this experience to his work on scientific communication.

Communication is the process of developing shared meaning. It shapes our experience of the world and of each other (Craig and Yewman, 2014). If we want our work to be taken seriously, we need to learn to be better communicators. This book is about how to do that.

The scientific process has the same structure as a story. The process begins when a scientist finds a problem with existing knowledge. She forms

a new hypothesis, generates data, and finally interprets the data and draws conclusions. This process matches the way stories have been told since time immemorial. A character is introduced in the context of the world as it exists (the current state of knowledge). Something happens (conflicting data) that causes them to question that world and changes their goals (the new hypothesis). They set out on an adventure (data collection, testing the hypothesis) with occasional setbacks (problems, unexpected results). Eventually they reach their goal (results) with new knowledge for the world (conclusions and significance). A new synthesis is reached. Existing knowledge is changed. The process begins again. It is a classic story played out in experiments and published in scientific journals. It is the hero's journey (Campbel, 1949). Let us begin with an example of scientific research and see how it fits this classic structure.

Dr. Pieter Visscher, Professor of Marine Sciences at the University of Connecticut, and his colleagues were interested in the oldest life on Earth: microbial communities that existed for over a billion years before there was oxygen in the atmosphere. These microbes secreted calcium carbonate and created layered limestone rocks called stromatolites. Because they existed billions of years ago, no one knew exactly how they captured energy from the sun. There was no oxygen, so they could not have used oxygenic photosynthesis or aerobic respiration to store and release energy. When Dr. Visscher's team found arsenic in fossil stromatolites they speculated that the ancient microbes may have used different kinds of arsenic as an electron donor in photosynthesis and an electron acceptor in respiration (Sforna et al., 2014). They wondered if similar modern-day communities also use arsenic, without the presence of oxygen. If they do, the team would have evidence for their theory. To test this idea, they looked for living stromatolites in the extreme conditions of the high Andes. In a small stream deep in the Atacama Desert they found bright purple microbial mats that thrive in the complete absence of oxygen and build stromatolites. Just as the clues they had found in ancient fossils suggested, these microbes used two different forms of arsenic to perform photosynthesis and respiration (Visscher et al., 2020b). Their discovery offers the strongest evidence yet for how the oldest life on Earth survived in a world without oxygen (Visscher *et al.*, 2020a). The fit with classic story structure is amazingly good.

Visscher *et al.* began their work in the familiar world where they knew that stromatolites existed, but they did not know how they captured energy. Based on their knowledge of the mechanisms of anoxygenic photosynthesis (current state of knowledge), they took their first step into the woods and investigated the composition of ancient stromatolites.

They found arsenic (conflicting data). Based on this they took a further step and speculated that the ancient microbes used arsenic as an electron donor in photosynthesis and acceptor in respiration (the new hypothesis). Now they had to gird themselves with the tools of modern chemistry and investigate the chemistry of contemporary stromatolite communities. If these microbes used arsenic for photosynthesis and respiration, the scientific adventurers would have evidence for their theory. To test this idea, they searched for living stromatolites in the extreme conditions of the high Andes (data collection, testing the hypothesis). Returning to their labs with samples collected under the most extreme conditions they found that these microbes used two different forms of arsenic (results). They returned from the woods with new knowledge (conclusions and significance). Their discovery changed our understanding of the oldest forms of life on Earth.

For Dr. Visscher and his team, and for every scientist who publishes their work, science is about changing the world (Besley, 2020). We do science to enrich our own understanding. We communicate our results to share that understanding with the world. As meaningful as the results are to us, we find that meaning heightened when we share them with others.

After the author had finished his first presentation at a major scientific meeting, one of the scientists in his field, someone whose work he had been reading during his studies, introduced herself and congratulated him on a nice presentation. While it felt good to pass his oral exam, turn in his dissertation, and graduate, there was nothing quite like the feeling of appreciation that came from this acknowledgement.

As scientists, we have learned to study nature; to ask and answer questions; to generate hypotheses, data, and results. Now we face a problem our science cannot solve. Society is increasingly bent on social media constructs. It does not take our work seriously. It does not seem to value truth. We live in a post-truth world where tribal membership seems more important than truth, where political affiliation overrules fact, and social media is used to manipulate belief. It is a time when truth is not just relative but manufactured to serve specific ends. We live in a society of instant gratification, one which seeks immediate answers to complex problems—answers that meet preconceived notions of how the world should be. We must, as scientists, as people who know that our knowledge has been hard-won and is closer to the truth than any other, be able to convey this knowledge to others in a post-truth world. This book is about how to do this. This book is about how to tell true stories.

Whether you are a student or a senior scientist, you can learn to be a better storyteller. This will make you a better communicator. It will help people understand your work. The skills required are a natural outgrowth of the scientific process and of the technical communication that is part of every scientist's work. In practicing them you will have the chance to integrate your analytic and artistic sides. All too often scientists have had to shut down their artistic, intuitive sides to complete their studies. When the author was applying to graduate school, he asked a question while being interviewed at one of his top candidate schools. He asked if it would be possible to continue his studies in other departments, such as the history of art. The answer was a definite no. His application was not accepted, and he never asked that question again. Although his question arose from a certain naïvety about the amount of work it would take to complete a PhD, it also reflected his desire to maintain a connection to the artistic side of his nature. Learning to tell true stories about your research is a way to maintain this connection. As Manasi Kulkarni-Khasnis writes, reconnecting with her interest in music helped her get through graduate school and complete her degree:

> Every composition I made came with a neurological boost. My experiments
> started working, or, more accurately, the failures weighed upon me less.
> I started to see each unexpected result as a new question to explore, rather
> than as a roadblock in my own work.
>
> (Kulkarni-Khasnis, 2018)

Nick Griffiths points out some of the parallels between doing science and playing music.

> Neither process is smooth. In science there are unexpected results, or no
> significant results at all, while in music there are technical difficulties that arise
> as fast as you can fix them. Also, they are both very solitary practices, though
> there are always colleagues and mentors around to help. But, it's only at the
> concert or conference that you, as a musician or scientist, finally share
> everything with the world.
>
> (Griffiths, 2015)

In the following chapters, you will learn how to present your science as a story, both in short verbal presentations and as a poster. You will learn the basic structure of stories. You will see how well this structure fits the scientific process. Using this structure will help you to improve your talks both to your colleagues and to the public. These presentations will be simple, surprising, concrete, emotional, stories (Heath and Heath, 2007).

This book will help you become a better communicator. It will help you present your research more clearly and with greater enthusiasm. It will help you write better grants, write better papers, and give better talks. It will help you present your ideas more clearly and with greater force. You will learn many techniques in this book. The techniques will improve your communication skills and help you connect with your audience. However, this book is not primarily about techniques. It is about expanding your potential as a human being. In learning to tell good stories you will find yourself, not just as a scientist but as an artist; as someone who can present their research in a way that is easily understood and has an impact on society. This book is about uncovering the parts of yourself that you had to shut down so that you could become a scientist. It is about reconnecting with the parts of yourself that are intimately connected with other people.

Science and the Media

Part of the problem with contemporary scientific communication is that many scientists have handed the task of explaining their work to the media. They live in a world of data and grants and allow others to decide what their work means for society. But the media misrepresents the nature of science. For the media, science is about results. The media approaches science as a series of questions and answers: How is the new virus transmitted? Will the comet pass close to the Earth? Do plants have feelings? The process of obtaining these results is of little interest to them. Very few scientists become journalists, so the science that makes it to the public has almost always passed through the filter of a person who is focused on answers. The process of science is lost. As good as some of the reporting is, the stories rarely capture the essence of the scientific process.

The disconnect between science as it is done, and science as it is presented has led to initiatives such as The Story Collider (https://www.story collider.org/), which helps people tell personal stories about science. These are not stories about data or results. They are personal stories about how science has impacted people's lives, and of the human experience behind scientific work. The Story Collider is the response to the data-driven presentations that most scientists give. There is a need for balance because the focus on data and results gives a one-sided view of science. It appears to many that science is done by smart, white-coated automatons who are focused on their data and are incapable of human interaction. This perception arose out of how science is presented, not how it is done. The practice of science and its presentation have been disconnected. The caricatures of science in the media are a direct outcome of this disconnect. Organizations like The Story Collider have arisen to address this problem.

Science has changed the world many times. From the discoveries of Galileo to the latest discoveries about dark matter, science has shaped how

we see the world and how we interact with it. John Snow's discovery of the waterborne transmission of cholera in 19th-century London (Snow, 1855), or Christiaan Eijkman's observations on beriberi that eventually led to the discovery of nutritional diseases and vitamin B_1 (Lindeboom, 2020) are classic examples of world-changing discoveries. But science has not always been well received. Great discoveries are often met with skepticism and hostility. The attacks on climate science and evolution are just the latest in a long series of controversies that have plagued scientific discoveries. Great discoveries challenge people's view of the world. For some, this is just not acceptable. They find security in the world-as-they-know-it and cannot face a new world where they do not know the rules.

The solution is not to rail against the anti-intellectualism that underlies this skepticism, but to learn to meet these needs by telling better stories. After all, this skepticism comes out of a deep desire for wish fulfillment. It is not based on data. It cannot be. Climate deniers do not need data because they know intuitively that the conclusions are wrong, so the data must be wrong. Deniers are focused on the story that scientists tell. It is the story to which they object. It is the story that must be told differently to meet these objections. If the data was going to convince them, then it would already have done so. The data on the anthropogenic origins of climate change is indisputable and has been so for over 30 years. But data does not meet the deep human need for stories. It especially does not meet the need for stories that have a redemptive ending. When well told, scientific stories are redemptive. They speak to the transcended experience of discovery and of the personal transformation that accompanies it (Shils, 1973). Each scientist is a hero who pushes the frontier of knowledge a bit farther into the darkness. Our presentations need to express this fact. We begin this process in the next chapter when we delve more deeply into the nature of stories and their relation to the scientific process.

Besley, J.C. (2020) Scientists don't share their findings for fun – they want their research to make a difference. *The Conversation*. The Conversation US, Boston, Massachusetts. Available at: https://theconversation.com/scientists-dont-share-their-findings-for-fun-they-want-their-research-to-make-a-difference-146267

Campbel, J. (1949) *The Hero with a Thousand Faces*, 1st edn. Pantheon Books, New York.

Craig, A. and Yewman, D. (2014) *Weekend Language: Presenting with More Stories and Less PowerPoint*. DASH Consulting, Portland, Oregon.

Griffiths, N. (2015) Musicians may make better scientists. *Helix Magazine*. Northwestern University: Science in Society. Available at: https://helix.northwestern.edu/blog/2015/03/musicians-may-make-better-scientists

Heath, C. and Heath, D. (2007) *Made to Stick*. Random House, New York.

Henderson, A. (1955) Henderson Recalls Shaw. *Durham Morning Herald*, August 21.

Kulkarni-Khasnis, M. (2018) How music helped me to flourish. *Nature Career Column*. Available at: https://www.nature.com/articles/d41586-018-05899-z

Lindeboom, G. (2020) *Eijkman, Christiaan*: Complete Dictionary of Scientific Biography, Encyclopedia.com. Available at: https://www.encyclopedia.com/people/medicine/medicine-biographies/christiaan-eijkman

Sforna, M.C., Philippot, P., Somogyi, A., van Zuilen, M.A., Medjoubi, K., Schoepp-Cothenet, B., Nitschke, W., and Visscher, P.T. (2014) Evidence for arsenic metabolism and cycling by microorganisms 2.7 billion years ago. *Nature Geoscience* 7, 811–815. doi: https://doi.org/10.1038/ngeo2276

Shils, E. (1973) The redemptive power of science. *Minerva* 11, 1–5.

Snow, J. (1855) *On the Mode of Communication of Cholera*. John Churchill, London.

Visscher, P., Brurns, B.P. and Gallagher, K.L. (2020a) Ancient microbial life used arsenic to thrive in a world without oxygen. *The Conversation*. The Conversation US, Boston, Massachusetts. Available at: https://theconversation.com/ancient-microbial-life-used-arsenic-to-thrive-in-a-world-without-oxygen-146533

Visscher, P.T., Gallagher, K.L., Bouton, A., Farias, M.E., Kurth, D., Sancho-Tomás, M., Philippot, P., Somogyi, A., Medjoubi, K., Vennin, E., Bourillot, R., Walter, M.R., Burns, B.P., Contreras, M., and Dupraz, C. (2020b) Modern arsenotrophic microbial mats provide an analogue for life in the anoxic Archean. *Communications Earth & Environment* 1: 24. doi: https://doi.org/10.1038/s43247-020-00025-2

A Deeper Look at Narrative Structure

2

The process of science has the same structure as a classic story. The hero leaves the familiar world, explores the unknown and returns with gifts that enrich their community. In scientific narratives, the familiar world is the current state of knowledge. It is the existing world in which the rules are known; the playing field is stable. The hero's departure is the new hypothesis and the tests to which it leads. They return with results that advance our

knowledge of the world. However, scientific process is seldom straightforward. Neither is the structure of a story. Both have twists and turns. Both have setbacks and unexpected results. There are backwaters and eddies that impede progress. The middle is usually tumultuous. It contains blind alleys. It produces surprising results. There are mountains that must be climbed.

Focusing our presentations solely on data gives the wrong impression. It misleads the audience into thinking that science is about results when it is really about process. Questions set the researcher on their path, leading them to their results. Questions occur at every step, with every new problem. Eventually they lead to an answer. Yet the answer always opens new vistas and leads to more questions. The world is changed. A new horizon opens with new mountains to be explored and new questions to be answered.

Three- and Five-act Structures

Let us see if we can use the three- and five-act structures so common in plays and movies to help us tell good stories about the scientific process (Yorke, 2014). A typical three-act play has the following structure. We have noted the role of each act in parentheses, with its role in the theater to the left of the equals sign and its role in a scientific story to the right.

Act 1: We are introduced to a flawed character and learn about their world (the preexisting world = current state of knowledge).

Act 2: The main character is confronted with a new situation (the inciting incident = conflicting evidence) that changes their view of the world (new outlook = new hypothesis) and challenges them to take action (the action of the play = data collection).

Act 3: The main character confronts the new situation and is forced to change to save the day (resolution = conclusions).

In *Star Wars*, these scenes play out in a typical hero's journey.

Act 1: The impatient, impetuous Luke Skywalker wants to escape his uncle's farm and become a space pilot (the preexisting world).

Act 2: The death of his uncle forces him off Tatooine (inciting incident) and begins his training with Obi-Wan Kenobi (new outlook). The Death Star is activated, Obi-Wan is killed, and the Death Star is turned on the rebel base (the action of the play).

Act 3: Luke must master his impatience, learn to trust the Force, and destroy the Death Star (resolution).

Adapting this form to scientific research we see that the three-act structure has a perfect parallel with the way that scientific research takes place.

Act 1: What is already known (current state of knowledge).

Act 2: New data questions the status quo (conflicting evidence); a new hypothesis is formed (new hypothesis); experiments are conducted (data collection).

Act 3: Resolution (conclusions).

Here is Dr. Visscher's work from the previous chapter presented in a three-act structure.

Act 1: **current state of knowledge**
    a. Photosynthetic bacterial mats existed in the ancient Earth and built stromatolites.
Act 2: **conflicting evidence; new hypothesis; data collection**
    a. Arsenic is found in the ancient stromatolites (**conflicting evidence**).
    b. Arsenic was the electron donor and acceptor in the ancient stromatolites (**new hypothesis**).
    c. Bacterial mats found in the high Andes contain the predicted types of arsenic (**data collection**).
Act 3: **conclusions**
    a. The ancient stromatolites cycled arsenic.

We see this same pattern in the five-act structure common in Shakespeare's plays. To create a five-act structure we break Act 2 into three parts, the new Acts 2–4, and add additional details (Yorke, 2014). The descriptions of the first and last acts may also be slightly modified.

Star Wars:

Act 1: The impatient, impetuous Luke Skywalker wants to escape his uncle's farm and become a space pilot (the preexisting world = current state of knowledge).

Act 2: The death of his uncle forces him off Tatooine (the inciting incident = conflicting evidence) and begins his training with Obi-Wan Kenobi (new outlook = new hypothesis).

Act 3: The Death Star is activated, and Alderaan is destroyed. Luke and his companions are captured and pulled into the Death Star (the action of the play = first presentation of results).

Act 4: Obi-Wan is killed. Luke escapes the Death Star, which follows him to the rebel base (more action = problems and their solution).

Act 5: Luke must master his impatience, learn to trust the Force, and destroy the Death Star (resolution = conclusions).

The important things here are not the specific events but the fact that there is a gradual transformation of the main character. He overcomes his flaw, usually only after great struggle, and receives his reward, fame in *Star Wars*. A similar transformation takes place in science. Both scientific knowledge and the scientist are transformed. This is most clearly seen in the transformation a graduate student undergoes in completing her research. However, it is true of all scientists. The reward for society is new knowledge. The reward for the scientist is her growth as a scientist and the satisfaction she receives from her discoveries.

Breaking the middle into three parts allows us to explore the scientific process in more detail. We can delve into some of the problems and their solutions that occur in every research study.

Act 1: What is already known (the current state of knowledge).

Act 2: Questions grow as evidence accumulates (conflicting evidence) and a new hypothesis is formed (new hypothesis).

Act 3: Experiments and results; experiments lead to new experiments or a revised hypothesis, which lead to more experiments and results (first presentation of results).

Act 4: Persistence in the face of difficulties; failed experiments; hard work; the normal problems with getting good measurements; there can be doubt about whether the research is on the right track, and the hypothesis may again be modified (problems and their solution).

Act 5: Resolution—everything comes together; the results are conclusive (or not, in the case of negative results) and conclusions are reached (conclusions).

Before we apply this model to Dr. Visscher's work, let us take a deeper look at his research and see some of the difficulties he faced when studying modern-day stromatolites. He knew that arsenic was a possible candidate for the electron donor in photosynthesis. Other candidates were hydrogen, ferrous iron and reduced sulfur. When he found arsenic in the ancient stromatolites, he hypothesized that the ancient microbes may have used arsenic as an electron donor in photosynthesis and an electron acceptor in respiration (Sforna *et al.*, 2014). To test this hypothesis he had to find microbial communities that were most like those that existed on the early Earth, then determine if they used arsenic in these ways. He eventually found these communities in the high Andes, but the path to this discovery was tortuous. He investigated many localities around the world before he was invited, by chance, to the Andes. A group of Argentinian scientists were curious about the origin of the carbonate sedimentary rocks in some of the high-altitude lakes in the Atacama Desert, which lies on the border between Argentina and Chile. They contacted him for advice and eventually invited him on their expeditions. They wanted to know if the sedimentary rocks in these lakes were being produced by aerobic bacteria. When they mentioned that they had seen stromatolite-like structures in other lakes, Dr. Visscher was hooked. He accepted their invitation. This began several years of exploring and testing high-altitude Andean lakes. At first, they were not even looking for arsenic as the main interest of the Argentinians was in oxygenic stromatolites. They published their first paper together on this subject (Farías *et al.*, 2014). When a post-trip analysis found lithium, arsenic, and sulfide in one of their water samples, Dr. Visscher was convinced that they had found a very special environment. He convinced the others that they had to go back. The lake had been incredibly hard to find. No roads led to it. They had to fly in and out. They wanted to show that the bacteria were anoxygenic in all seasons, so a single return trip was not enough. In the end, they made six trips. During the winter, the night-time temperatures dropped below −20°C. In the summer it reached 40°C. Equipment froze and overheated. Their eyes watered for hours on end due to the extremely low humidity. Their skin burned because the ultraviolet radiation was 25 times higher than at sea level. They could not use normal oxygen microelectrodes because these sensors cannot measure below about 0.1% oxygen. Working with the Danish company Unisense (https://www.unisense.com/) they developed a new oxygen sensor that was over 50 times more sensitive. To convince their peers of the accuracy of their results they conducted lab experiments with slurries of the bacteria and were able to show that they use two different forms of arsenic to perform photosynthesis and respiration (Visscher *et al.*, 2020).

As with all research, there were plenty of problems. Let us put some of them into the five-act structure to see if we can tell a story that will hold the interest of an audience.

Act 1: current state of knowledge
    a.   Dr. Visscher knew that photosynthetic, carbonate-producing bacterial mats existed in the ancient Earth and built stromatolites.
    b.   Possible electron donors for anoxygenic photosynthesis included hydrogen, ferrous iron, reduced sulfur, and arsenic.
    c.   But what type of photosynthesis did they use? Which compound did they use as their source of electrons?

Act 2: conflicting evidence; new hypothesis
    a.   Arsenic is found in the ancient stromatolites.
    b.   A new hypothesis is formed: ancient stromatolites used arsenic as an electron donor for photosynthesis and an electron acceptor for respiration.

Act 3: first presentation of results
    a.   Dr. Visscher visits possible study sites in the Andes but does not find anoxygenic conditions.
    b.   He collects information on oxygenic bacterial mats and publishes a paper on oxygenic stromatolites with his Argentinean colleagues.

Act 4: problems and their solution
    a.   They find and visit a lake in the Atacama Desert and collect water samples for later analysis. The samples come back showing the presence of arsenic.
    b.   He wants to go back and take measurements, but existing oxygen sensors are not sensitive enough. A new oxygen sensor is developed.

    c. Oxygen measurements are made under the most extreme conditions imaginable.

    d. Oxygen is absent at all times of the year and the correct types of arsenic are found in the bacterial mats.

    e. Lab experiments show that arsenic cycling occurs in anoxic slurries of the bacteria.

Act 5: conclusions

    a. There is support for the use of arsenic by the ancient stromatolites.

    b. Because arsenic is cycled between photosynthesis and respiration, it could have supported ancient ecosystems for long periods of time.

The fit is amazingly accurate. Of course, we have chosen only some of the problems to present, but this is the same thing we have done in summarizing *Star Wars*. Princess Leia is a main character in *Star Wars* but is not mentioned in our summary. The point of the story is not to capture every event, but to summarize the main points in a way that will hold an audience's attention.

By framing the scientific process in this way, we tell stories that are simple, surprising, concrete, and emotional. These are the stories that stick (Heath and Heath, 2007). They are simple because they follow a pattern that is well established in people's minds based on their experience with movies and plays. They are surprising because new results are always surprising. They are concrete because they provide details about the scientific process. They do not jump between the current state of knowledge, the new hypothesis, and the answer. They provide details about how science is done. They are emotional because they allow the investigator to express her personal attachment to her work. They allow her to show that she cares about each step of the process. She cares so much that she is willing to undergo great hardship in the pursuit of her research. Audiences will identify with this. Because they identify with the process and with the investigator, they will take her results more seriously. These types of stories are impactful and will be remembered.

This is not to say that all science must be presented in a three- or five-act structure. However, these structures allow us to understand the scientific process in a way that strengthens our presentations. Short presentations of 5–10 minutes will more commonly follow a three-act structure while the five-act structure is more suitable for talks of up to 20 minutes. We will return to longer presentations in Chapter 5 where we will see how extensions of the five-act structure can be used for presentations of up to an hour.

Let us return to the three-act structure and consider another example.

Neurotoxic peptides from the giant Australian stinging tree

Dr. Irina Vetter, Associate Professor at the University of Queensland, and her colleagues use plant and animal toxins to understand the molecular pharmacology of pain. Given this interest, it was natural for her to work on the stinging trees of the Australasian genus *Dendrocnide* (Urticaceae), which produce painful and persistent stings. The compounds responsible for these stings reside

in the epidermal hairs (trichomes) that cover the surface of the plant. The pain usually lasts for several hours, but painful flares can continue for days, and sometimes weeks.

Based on what she knew about the pharmacologically active compound in the trichomes of the related stinging nettle (*Urtica*, Urticaceae) it was reasonable for Dr. Vetter to assume that this compound, a small molecular neuropeptide called moroidin, would also be found in the trichomes of *Dendrocnide* spp. However, human tests with purified extracts of moroidin had previously shown that it was unlikely to be the main culprit in *Dendrocnide* stings. When moroidin was injected alone, it failed to produce the substantial long-lasting pain that occurs with contact with the plant. Also, when her team tried to isolate moroidin from *Dendrocnide excelsa*, they found that it was not present. These results suggested that there were unidentified pharmacologically active compounds in *Dendrocnide* trichomes.

To test this hypothesis, they isolated and purified the molecular constituents from *D. excelsa* trichomes and assessed the sensory effects of the resulting fractions. Mass spectrometry of a late-eluting fraction showed the presence of miniproteins stabilized by disulfide bonds. Experiments with synthetic versions of these peptides showed that they can reproduce the symptoms of a *Dendrocnide* sting. The severity of these reactions suggested that the peptides might be neurotoxins. Patch-clamp electrophysiology of mouse dorsal root ganglion neurons confirmed that the peptides work directly on voltage-gated sodium (Na_V) channels, confirming this hypothesis.

Though this class of neurotoxins were previously unknown from plants, they are similar in structure to neurotoxins found in spider and cone snail venoms. Understanding the mechanisms by which these molecules affect ion channels and nerve cells may provide new ways of controlling pain.

Let us put this research into a three-act structure.

Act 1: current state of knowledge
    a.   Contact with *Dendrocnide* plants causes excruciating long-lasting pain.

Act 2: conflicting evidence; new hypothesis; data collection
    a.   Purified extracts of moroidin, the compound responsible for stings in related plants, do not produce the same long-lasting effects as *Dendrocnide* stings. *D. excelsa* trichomes lack moroidin (conflicting evidence).
    b.   Other pharmacologically active compounds present in the *Dendrocnide* trichomes must be responsible for the stings (new hypothesis).
    c.   The symptoms of a *Dendrocnide* sting can be reproduced from synthetic versions of disulfide-rich peptides isolated from the trichomes; patch-clamp electrophysiology demonstrates that these compounds are neurotoxins (data collection).

Act 3: conclusions
    a.   The causative agents of *Dendrocnide* stings are neurotoxins similar to those found in spider and cone snail venom.

The And-But-Therefore Model

Another way of looking at story structure is the And-But-Therefore (ABT) model popularized by marine biologist turned filmmaker Randy Olson (Olson, 2020). The And-But-Therefore model is similar to the three-act model,

but it places emphasis on different parts of the story. AND is equivalent to Act 1. It is the part of the story where we learn about the preexisting world. In the theater this is called the exposition. In scientific stories the exposition is about the current state of knowledge. It is the part of the presentation where we describe what was known before we began our work. It is the description of scientific knowledge before our new hypothesis. In Dr. Visscher's work it is the knowledge of possible electron donors in anoxygenic photosynthesis. In Dr. Vetter's study it is the fact that contact with *Dendrocnide* plants causes excruciating pain. The exposition is about the past. It describes the world without conflicting evidence.

In its simplest form BUT is a description of the conflicting evidence that leads to the new hypothesis. In the theater BUT is called the inciting incident. It is the event that changes the world. It is the death of Luke's uncle in *Star Wars*. It is the event that initiates change. In scientific stories BUT is the evidence that shows the need for a new hypothesis. It shows the problems with existing knowledge and demonstrates the need for new work. In a three-act story BUT occurs at the beginning of the second act. It prepares the audience for the new hypothesis. In Dr. Visscher's work it is the discovery of arsenic in the ancient stromatolites. In Dr. Vetter's *Dendrocnide* study it is the discovery that extracts of *D. excelsa* trichomes lack moroidin.

The world changes because of BUT. There are consequences. THEREFORE is an exploration of these consequences. In the theater, THEREFORE makes up the majority of the play. It chronicles the events that occur after the inciting incident. There are twists and turns, ups and downs. There are unexpected events that must be reconciled. Luke's path from Tatooine to his confrontation with the Death Star is tortuous. He meets and takes passage with Han Solo, is trapped in the Death Star, meets Princess Leia, watches Obi-Wan die, escapes the Death Star, is pursued by the Empire to the rebel base, and only then is in a position for the final confrontation.

In scientific presentations, THEREFORE begins with a statement of the new hypothesis, includes a description of the experiments and their results, and ends with the conclusions. The body of THEREFORE chronicles what happens because of the new hypothesis. What type of data is collected? What problems occur? What are the results? What conclusions are drawn, and what are their implications?

The And-But-Therefore structure emphasizes transitions. The three- and five-act structures emphasize our knowledge before, during, and after the new hypothesis. The transitions are present in these forms but are given less attention. The And-But-Therefore form forces us to recognize the importance of the transitions: the conflicting evidence (BUT) and how it leads to the new hypothesis (THEREFORE). These transitions are the keys to effective scientific storytelling.

Using the And-But-Therefore structure we can produce very short presentations, as we will see in the chapter on elevator pitches (Chapter 4). We must give up some of the details of our experiments to do this. For very short presentations these details are less important than the overall flow of the story. The transitions give us this flow. For use in short presentations, we can think of the structure in this way.

And: current state of knowledge
But: conflicting evidence
Therefore: new hypothesis, data collection, conclusions

Stories constructed with this form are short, simple, and surprising. The simplicity is inherent in the form. There are three simple parts: (i) current knowledge; (ii) conflicting evidence; and (iii) new hypothesis and results. The presentations are surprising partly because all research that overturns existing knowledge is surprising, and partly because the And-But-Therefore structure emphasizes transitions. BUT is surprising because it shows that our knowledge is incomplete and suggests the need for new work. THEREFORE is surprising because it provides a new understanding of something we thought we understood.

Let us now apply this structure to both Dr. Visscher's and Dr. Vetter's studies.

Ancient stromatolites and arsenic cycling

And: current state of knowledge
   a. Photosynthetic bacterial mats existed in the ancient Earth and built stromatolites.
   b. Possible electron donors for anoxygenic photosynthesis include hydrogen, ferrous iron, reduced sulfur, and arsenic.
But: conflicting evidence
   a. Arsenic is found in the ancient stromatolites.
Therefore: new hypothesis, data collection, conclusions
   a. Arsenic was an electron donor and acceptor in the ancient stromatolites (new hypothesis).
   b. Bacterial mats found in the high Andes contain the predicted forms of arsenic and slurries of bacteria from these mats cycle arsenic (data collection).
   c. The ancient stromatolites cycled arsenic (conclusions).

Neurotoxic peptides from the giant Australian stinging tree

And: current state of knowledge
   a. Contact with *Dendrocnide* plants causes excruciating, long-lasting pain.
But: conflicting evidence
   a. Purified extracts of moroidin do not produce the same long-lasting pain. *D. excelsa* trichomes lack moroidin.
Therefore: new hypothesis, data collection, conclusions
   a. Other pharmacologically active compounds are present in the trichomes (new hypothesis).
   b. The symptoms of a sting can be reproduced from synthetic versions of disulfide-rich peptides isolated from the trichomes (data collection).
   c. The causative agents are neurotoxins similar to those found in spider and cone snail venom (conclusions).

The problem with the And-But-Therefore structure is that the conflicting evidence is a relatively minor part of longer presentations. Most of a longer talk will be devoted to new experiments and results. The conflicting evidence only serves to set the stage. In very short talks, the conflicting evidence plays a much larger role. It is the hook that gets the audience interested. It tells them that something is wrong with the world. It tells them that they should pay attention. There is much less time to present our results in very short presentations. This difference between very short, slightly longer, and long presentations is the reason we have presented three different models. The And-But-Therefore model is best suited to very short presentations of 90 seconds or less.

The Landscape of Short Presentations

For the purposes of this book, we will consider any verbal presentation shorter than 15 minutes a short presentation. At the shorter end of the spectrum are elevator pitches and the many forms of lightning talks. At the longer end there are 10–15-minute conference presentations. Longer talks are those of half an hour to an hour. Poster presentations fall somewhere between elevator pitches and short presentations. They can be as short as a few seconds or as long as a conference talk, sometimes longer if someone is interested in your work.

The first recorded use of short talks in a conference format was at the Sixth International Python Conference held in San Jose, California (USA) in October of 1997. What would later be called Lightning Talks began as 10-minute talks intended to present breaking research in a short format (Anonymous, 1997). The talks themselves were 7 minutes, after which the speaker was cut off and 3 minutes were left for ques-

− Lightning Talks −

tions. No slides were permitted. The submission requirements were less rigorous than for standard talks, and the submission deadline was considerably later. Lightning talks have been part of every subsequent Python conference and have spread to many other disciplines. Unlike many of their later derivatives the original lightning talks were technical talks aimed at a professional audience.

A number of other short-talk formats have been developed in subsequent years. Almost all of them are aimed at the general public. Some of the more prominent forms are:

- Flash Talks (3 minutes, no slides, props allowed) (Appiah, 2020)
- Three Minute Thesis (3MT®, 3 minutes, one slide, no props) (The University of Queensland, 2008-present)
- Three Minute Wonder (3 minutes, one slide or video, props allowed) (Institute of Physics, 2021)
- FameLab (3 minutes, no slides, props allowed) (British Council, 2021)
- Perfect Pitch (90 seconds, one slide) (Engineering Research Centers, 2021)

There are also other forms of short talks hosted by individual universities, and even by individual departments. The forms of these talks are usually variations on one of those described above. A number of scientific societies have also begun incorporating short, usually student presentations in their conferences. These talks are most often aimed at a technical not a general audience.

On the comedic side of presentations aimed at a general audience there is the Science Showoff and related forms like Bright Club in the UK and Australia (Science Showoff, 2011-present), and Stand Up Science (Mauss, 2021), which has had various incarnations in the USA. Sitting right between the comedic and more sober presentations to general audiences we find Ignite® talks (5 minutes, 20 slides for 15 seconds each) (Ignite Talks PBC, 2006–2021) and PechaKucha™ (20 slides for 20 seconds each) (PechaKucha, 2021). PechaKucha talks can be on any subject. They are not restricted to science. All of these events combine science and comedy and are beyond the scope of this book.

Your Turn

This is an interactive book. We have interspersed exercises throughout so you can practice the skills you need to become a better communicator. You will get the most out of this book if you practice these exercises so you can gain experience with the forms. There follow three descriptions of research for your analysis. Read them carefully, then put them into three-act and And-But-Therefore structures. Our formulations are at the end of the chapter, but do not look at them until you have completed your own summaries.

Male sex determination in mice

Sry is a gene on the Y chromosome of mice that is responsible for the formation of testes, and therefore for the production of males (Gubbay *et al.*, 1990). The sex chromosome complement of males is XY. Females are XX, and so lack *Sry*. Since its discovery in 1990 scientists have thought that *Sry* consisted of a single contiguous piece of DNA that codes for one protein (a single exon). This finding was called into question when Dr. Makoto Tachibana's lab at Osaka University used RNA sequencing (RNA-Seq) to search for other genes involved in male sex determination (Wang *et al.*, 2009). To verify that their methods were working correctly they examined *Sry* expression expecting to find a single RNA transcript: the one from *Sry*. Instead, they found an unknown transcript, related to *Sry* but mapping to a different position on the chromosome. This appeared to be a second copy of the *Sry* gene (Miyawaki *et al.*, 2020). If verified, this meant that there were two exons potentially involved in male sex determination. The discovery of the second exon was very surprising. The mouse genome has been extensively studied and geneticists assumed that all of the genes, and all of their parts, had been found, even if their functions had not been fully characterized. The discovery of this second exon led to the hypothesis that it was also involved in male sex determination. To test this, Tachibana's lab used the gene editing technology CRISPR (Clustered Regularly Interspaced Short Palindromic Repeats) to remove the new exon from the mouse genome. This new strain of mice was able to make the protein encoded by the original exon but lacked the protein encoded by the new exon. Even though these mice possessed a Y chromosome and made the original Sry protein, they did not develop testes. They were female. This demonstrated that the original exon was not sufficient to produce males. To further check this result, Tachibana's lab performed the reverse experiment: they used knockout technology to express the protein encoded by the new exon in an XX genetic background (Hansson, 2007). Based on their sex chromosomes, these mice should have been female. They were not. They were male.

In working to understand these results, Tachibana's lab performed experiments to uncover the function of the new exon. Analysis of the protein encoded by the original exon showed that it has a motif inducing rapid degradation at one end, a degron. The protein unravels like a sweater

without a hem and cannot function. The protein encoded by the new exon does not have a degron. This protein survives and functions to direct normal male development. These findings explain why removing the new exon from XY mice causes female development. The remaining protein from the first exon degrades and cannot carry out its function. It also explains what happens when the protein encoded by the new exon is expressed in XX mice: a properly protected protein is made and can direct testis development. The mice become male.

Now put this research into three-act and And-But-Therefore structures. Our answers are at the end of the chapter.

Restoration of seagrass habitat in coastal Virginia

A century ago, in the coastal waters and inshore lagoons in the state of Virginia (USA), beds of eelgrass (*Zostera marina*) provided habitat for bay scallops (*Argopecten irradians*), for the Brent goose (*Branta bernicla*), and stabilized the sea floor. Eelgrass was so common that people who lived near the shore baled it to use as insulation for their homes. In the 1920s, the marine slime mold *Labyrinthula zosterae* (eelgrass blight) began wiping out the beds so that by the mid-1930s the plant was virtually extinct in the Virginia lagoons. With its demise, bay scallops and the Brent goose disappeared, along with other organisms that depend on eelgrass. With the disappearance of eelgrass water turbidity increased, which led to the hypothesis that decreased light quality was responsible for the lack of eelgrass recovery. However, in the late 1990s a natural patch of eelgrass was discovered by a local waterman, probably from seeds hitchhiking on recreational boats. Subsequent studies showed that light was not the limiting factor preventing spread of the plants.

Based on these observations, teams lead by Dr. Robert Orth at the Virginia Institute of Marine Science and Dr. Karen McGlathery at the University of Virginia hypothesized that the lack of recovery of eelgrass was due to the lack of a seed source rather than light quality (turbidity). To test this hypothesis, they established test plots of eelgrass in the lagoons. Between 2001 and 2004, approximately 24 million seeds were harvested from Chesapeake Bay, just east of the lagoons, and broadcast into experimental plots in four coastal lagoons having no eelgrass. Low-level aerial photographs in 2004 showed 38% of the area of the plots was covered by vegetation. These plots demonstrated that seed supply was the limiting factor in eelgrass recovery (Orth *et al.*, 2006). The team then enlisted volunteers to broadcast seeds into these areas. Over the next 16 years volunteers spread more than 74 million seeds into 536 restoration plots totaling 213 ha. Long-term monitoring of the restored beds showed that 3612 ha of vegetated bottom has been restored.

A restoration program for bay scallops was begun in 2008. Annual seeding efforts have resulted in a sustained population of scallops as revealed by the presence of scallops in aquaculture beds up to 20 km away from where they were set out. The program has restored a remarkably hardy ecosystem that is trapping carbon and nitrogen; carbon that would otherwise contribute to global warming and nitrogen that would otherwise diminish water quality. It is one of the most successful marine restoration efforts in the world (McGlathery *et al.*, 2012; Orth *et al.*, 2020).

It is now your turn to put this research into three-act and And-But-Therefore structures. Jump to the end of the chapter to see our formulations.

Uncovering the origins of the sandstone megaliths at Stonehenge

Stonehenge, the Neolithic stone circle in south-west England, was constructed around 2500 BCE from stones from at least three source areas. The locations of these areas have been debated for over four centuries. The smaller igneous and sandstone "bluestones" used near the center of the monument have been traced to two different localities in Wales. The origin of the sandstone megaliths that form the massive uprights and lintels of the primary architecture have traditionally been thought to originate from the Marlborough Downs (downs: from the Old English *dūn*, meaning hill), the closest area with a substantial supply of large sandstone boulders of the correct type. However, this supposition has never been scientifically verified. The Marlborough Downs cover an area of over 30 km², so narrowing the providence of the stones and testing the conventional wisdom has been of interest to geographers and archeologists for many years.

Work on the stones has long been difficult because of the inability to destructively sample from a site of such immense cultural significance. This situation changed in 2018 when a meter-long drill core from one of the sandstone megaliths (Stone 58) was returned to English Heritage. Three drill cores were taken from Stone 58 during conservation work in 1958 and had been presumed lost. The return of one of the missing cores allowed a team of scientists led by Professor David Nash at the University of Brighton, with the permission of English Heritage, to destructively sample a piece of stone from the unweathered center of a megalith (Nash *et al.*, 2020). Before undertaking this work, Professor Nash's team had used non-destructive X-ray fluorescence spectrometry to investigate the chemical composition of the 52 remaining megaliths at the monument. The other 28 original megaliths had been removed in antiquity. Linear discriminant analysis of the X-ray data coupled with Bayesian principal component plots demonstrated that 50 of the 52 megaliths, including Stone 58, shared a consistent geochemistry and likely originated from a common source. Once this was established, the team used inductively coupled plasma mass spectrometry to destructively sample a small portion of the core from Stone 58 (Nash *et al.*, 2013). They compared the resulting geochemical signature with equivalent data for similar stones from across southern Britain. From this, they identified an area 25 km north of Stonehenge (West Woods, at the south-eastern edge of the Marlborough Downs), as the most probable source for the majority of the megaliths at the monument. The providence of the two megaliths with different geochemical signatures remains a mystery.

Now put this research into a three-act structure. There is no conflicting evidence in this case, so you will have to use your ingenuity to construct the second act. The event that made the research possible was the return of the drill core from Stone 58. This made it possible to conduct destructive tests. Because there is no conflicting evidence it is difficult to use the And-But-Therefore form to describe this research. Not all research will fit well into every format. That is one reason we present multiple ways to construct a scientific story. In this case the And-But-Therefore format is not a good fit. Our answer is at the end of the chapter.

Answers to Exercises

Male sex determination in mice

Three-act structure

Act 1: **current state of knowledge**
  a. The *Sry* gene has a single exon and is responsible for male sex determination in mice.

Act 2: **conflicting evidence; a new hypothesis; data collection**
  a. New sequencing techniques lead to the discovery of a second exon (**conflicting evidence**).
  b. The new exon is needed for the formation of testes in mice (**new hypothesis**).
  c. Removing the new exon causes XY mice to be female; expressing it in XX mice causes them to be male (**data collection**).

Act 3: **conclusions**
  a. *Sry* has two exons. One produces a protein that rapidly degrades. The other produces a functional protein.

And-But-Therefore structure

And: **current state of knowledge**
  a. The *Sry* gene has a single exon and is responsible for male sex determination in mice.

But: **conflicting evidence**
  a. A new exon of the gene is discovered.

Therefore: **the new hypothesis, data collection, conclusions**
  a. The new exon is needed for the formation of testes in mice (**new hypothesis**).
  b. Removing the new exon causes XY mice to be female; expressing the new exon in XX mice causes them to be male (**data collection**).
  c. The new exon is necessary for male sex determination in mice (**conclusions**).

Restoration of seagrass habitat in coastal Virginia

Three-act structure

Act 1: **current state of knowledge**
  a. Eelgrass blight wiped out eelgrass in the Virginia lagoons. With its disappearance water turbidity increased. Increased turbidity led to the hypothesis that decreased water quality was responsible for the lack of eelgrass recovery.

Act 2: **conflicting evidence; new hypothesis; data collection**
  a. Reappearance of eelgrass in one of the lagoons and studies that showed light was not the limiting factor called this hypothesis into question (**conflicting evidence**).
  b. Lack of seed rather than water quality is responsible for the lack of recovery in the lagoons (**new hypothesis**).

Continued

Continued

> c. Test plots were seeded and established spreading colonies of eelgrass. Lack of seed was thereby shown to be the limiting factor (**data collection**).
>
> Act 3: **conclusions**
>
> a. A reseeding project was established and spread more than 74 million seeds over 213 ha of the southern lagoons. A restoration program for bay scallops was also started. The project has restored over 3500 ha of the lagoons to eelgrass beds. Bay scallops have been found in aquaculture beds up to 20 km away from where they were set out.
>
> *And-But-Therefore structure*
>
> And: **current state of knowledge**
>
> a. Eelgrass blight wiped out eelgrass in the Virginia lagoons. With its disappearance water turbidity increased. Increased turbidity led to the hypothesis that decreased water quality was responsible for the lack of eelgrass recovery.
>
> But: **conflicting evidence**
>
> a. Eelgrass recovery occurred in one of the lagoons.
> b. Studies showed light was not the limiting factor.
>
> Therefore: **new hypothesis, data collection, conclusions**
>
> a. Lack of seed rather than water quality is responsible for the lack of eelgrass recovery (**new hypothesis**).
> b. Test plots sown with eelgrass seed demonstrate that restoration can occur in the lagoons (**data collection**).
> c. Long-term reseeding of the lagoons has restored over 3500 ha of eelgrass. A successful restoration program for bay scallops was started (**conclusions**).

> ## Uncovering the origins of the sandstone megaliths at Stonehenge
>
> *Three-act structure*
>
> Act 1: **current state of knowledge**
>
> a. The sandstone megaliths at Stonehenge are thought to have originated from the Marlborough Downs, but there has been no scientific test of this hypothesis.
>
> Act 2: **conflicting evidence; new hypothesis; data collection**
>
> a. A meter-long drill core from Stone 58 is returned to English Heritage (**conflicting evidence = the event that made the research possible**).
> b. A test of the Marlborough Downs hypothesis is now possible (**new hypothesis**).
> c. X-ray fluorescence spectrometry shows that 50 of the 52 remaining megaliths came from a single source area.
> d. inductively coupled plasma mass spectrometry is used to compare Stone 58 with similar data from areas across southern Britain (**data collection**).
>
> Act 3: **conclusions**
>
> a. The data localizes Stone 58, and by association 49 other megaliths, to an area on the south-eastern edge of the Marlborough Downs.

Anonymous (1997) Wanted: Short Talks. The Sixth International Python Conference, San Jose, California. Available at: https://legacy.python.org/workshops/1997-10/shorties.html

Appiah, B. (2020) How to give a science flash talk. Script Practical Guide. Script. Available at: https://scripttraining.net/script-practical-guide/media-skills-for-scientists/how-to-give-a-science-flash-talk/

British Council (2021) FameLab: Science Communication Competition. Available at: https://www.britishcouncil.org/education/he-science/famelab

Engineering Research Centers (2021) Perfect Pitch Guidelines. National Science Foundation's (NSF) Engineering Research Centers, USA. Available at: https://erc-assoc.org/content/perfect-pitch-guidelines

Farías, M.E., Contreras, M., Rasuk, M.C., Kurth, D., Flores, M.R., Poiré, D.G., Novoa, F., and Visscher, P.T. (2014) Characterization of bacterial diversity associated with microbial mats, gypsum evaporites and carbonate microbialites in thalassic wetlands: Tebenquiche and La Brava, Salar de Atacama, Chile. *Extremophiles* 18, 311–329. doi: https://doi.org/10.1007/s00792-013-0617-6

Gubbay, J., Collignon, J., Koopman, P., Capel, B., Economou, A., Münsterberg, A., Vivian, N., Goodfellow, P., and Lovell-Badge, R. (1990) A gene mapping to the sex-determining region of the mouse Y chromosome is a member of a novel family of embryonically expressed genes. *Nature* 346, 245–250. doi: https://doi.org/10.1038/346245a0

Hansson, G.K. (2007) *Gene Modification in Mice*. The Nobel Prize in Physiology or Medicine 2007 – Advanced Information. Nobelförsamlingen, Stockholm. Available at: https://www.nobelprize.org/prizes/medicine/2007/advanced-information/

Heath, C. and Heath, D. (2007) *Made to Stick*. Random House, New York.

Ignite Talks PBC (2006-present) *Ignite®*. *Ignite is a series of speedy presentations*. Available at: http://www.ignitetalks.io/

Institute of Physics (2021) Three Minute Wonder Science Communication Competition. Institute of Physics (IOP), UK. Available at: https://www.iop.org/3mw

Mauss, S. (2021) Stand Up Science. Available at: https://www.shanemauss.com/science

McGlathery, K.J., Reynolds, L.K., Cole, L.W., Orth, R.J., Marion, S.R. and Schwarzschild, A. (2012) Recovery trajectories during state change from bare sediment to eelgrass dominance. *Marine Ecology Progress Series* 448, 209–222.

Miyawaki, S., Kuroki, S., Maeda, R., Okashita, N., Koopman, P. and Tachibana, M. (2020) The mouse *Sry* locus harbors a cryptic exon that is essential for male sex determination. *Science* 370, 121–124. doi: https://doi.org/10.1126/science.abb6430

Nash, D.J., Coulson, S., Staurset, S., Ullyott, J.S., Babutsi, M., *et al.* (2013) Provenancing of silcrete raw materials indicates long-distance transport to Tsodilo Hills, Botswana, during the Middle Stone Age. *Journal of Human Evolution* 64, 280–288. doi: https://doi.org/10.1016/j.jhevol.2013.01.010

Nash, D.J., Ciborowski, T.J.R., Ullyott, J.S., Pearson, M.P., Darvill, T., *et al.* (2020) Origins of the sarsen megaliths at Stonehenge. *Science Advances* 6: eabc0133. doi: https://doi.org/10.1126/sciadv.abc0133

Olson, R. (2020) *The Narrative Gym*. Prairie Starfish Press, Los Angeles, California.

Orth, R.J., Luckenbach, M.L., Marion, S.R., Moore, K.A. and Wilcox, D.J. (2006) Seagrass recovery in the Delmarva Coastal Bays, USA. *Aquatic Botany* 84, 26–36. doi: https://doi.org/10.1016/j.aquabot.2005.07.007

Orth, R.J., Lefcheck, J.S., McGlathery, K., Aoki, L., Luckenbach, M.W., *et al.* (2020) Restoration of seagrass habitat leads to rapid recovery of coastal ecosystem services. *Science Advances* 6: eabc6434. doi: https://doi.org/10.1126/sciadv.abc6434

PechaKucha (2021) PechaKucha: Visual Storytelling that Celebrates Humanity. Available at: https://www.pechakucha.com/

Science Showoff (2011-present) Science Showoff. Available at: http://www.scienceshowoff.org/

Sforna, M.C., Philippot, P., Somogyi, A., van Zuilen, M.A., Medjoubi, K., *et al.* (2014) Evidence for arsenic metabolism and cycling by microorganisms 2.7 billion years ago. *Nature Geoscience* 7, 811–815. doi: https://doi.org/10.1038/ngeo2276

The University of Queensland (2008-present) *3MT®* Three Minute Thesis. The University of Queensland, Brisbane, Queensland, Australia. Available at: https://threeminutethesis.uq.edu.au/

Visscher, P.T., Gallagher, K.L., Bouton, A., Farias, M.E., Kurth, D., Sancho-Tomás, M., Philippot, P., Somogyi, A., Medjoubi, K., Vennin, E., Bourillot, R., Walter, M.,R., Burns, B.P., Contreras, M., and Dupraz, C. (2020) Modern arsenotrophic microbial mats provide an analogue for life in the anoxic Archean. *Communications Earth & Environment* 1: 24. doi: https://doi.org/10.1038/s43247-020-00025-2

Wang, Z., Gerstein, M. and Snyder, M. (2009) RNA-Seq: a revolutionary tool for transcriptomics. *Nature Reviews Genetics* 10, 57–63. doi: https://doi.org/10.1038/nrg2484

Yorke, J. (2014) *Into the Woods: a Five-Act Journey Into Story*. Abrams Press, New York, p. 336.

Presenting in Three Minutes

The path a scientist pursues in their research begins with existing but flawed knowledge, proceeds through a series of explorations and culminates in new knowledge. In the previous chapters we saw how the discovery of arsenic in the ancient stromatolites set Dr. Visscher's team off in a new direction, and how the failure to find moroidin in *Dendrocnide excelsa* caused Dr. Vetter to look closer at the pharmacological constituents of the trichomes. These inciting incidents led the research in new directions. Eventually, after many observations and experiments, they led to new understandings. Our view of the world changed.

The path that a graduate student takes from their first semester to their dissertation defense follows a similar track. A student enters a new field with confusing terminology and strange equipment. They must vanquish their insecurities, learn new techniques, and learn a new language so that they can take their place as a master. They leave having completed a project that they could barely imagine when they entered. In the process they have had to overcome their self-doubts, learn new ways of behaving, master arcane techniques and terminology, and transform themselves into scientists. It is the classic hero's journey. It is a path of transformation. It is also a path of discovery.

These two aspects of scientific stories, the story of the research, and the story of your personal transformation can be effectively combined to present your research in 3 minutes. Giving your audience a feeling for your personal journey will increase your connection and help them understand your research. We have all had to sit through boring, dispassionate presentations that were hard

to understand and left us disconnected from the speaker. You can avoid this problem by explaining why your research is important to you and by being present for your audience. If you feel passionately about your work, you can show your audience this passion as Dr. Joshua Chu-Tan does in his 3-minute talk, reviewed later in this chapter. If you began your research because of a personal experience, you can briefly explain this as Dr. Yasmin Mustapha Kamil does in her talk. You do not have to go into great detail. For now, it is enough to realize that there is a personal element to all research and that this element should be part of your presentation.

Presenting a Three Minute Thesis

There are many types of 3-minute presentations. Since they share many similarities, we will choose one to start and discuss some of the variations later.

We will start with the Three Minute Thesis (3MT®). The 3MT was developed and copyrighted by The University of Queensland (2008-present). As the title implies, a Three Minute Thesis is a verbal presentation of masters or dissertation research in 3 minutes. A single static PowerPoint slide is permitted, but there can be no props or additional electronic media such as audio or video files. The judging criteria include content and engagement components[1]. Institutions that want to host a 3MT must receive permission from the university, must use the official judging criteria, and must adhere to the contest rules. Each presentation is sometimes, but not always, followed by a question-and-answer period. When present, the question period is not judged.

Aside from presentation style and audience connection, which we deal with in later chapters, there are two important things for you to consider in preparing your 3MT presentation: (i) the organization of your talk; and (ii) the relationship of your presentation to your slide. While a slide is not required in a 3MT, a slide can greatly enhance your presentation. We recommend using one. Every 3MT we have seen has used a slide. We will consider how you can present well without a slide when we talk about FameLab presentations later in this chapter. The rules for FameLab do not allow slides, but props are permitted (British Council, 2021).

The easiest way to think of a 3MT is as a very short paper presented to a general audience. Every scientific paper has these parts: introduction, methods, results, and discussion. The discussion normally includes your overall conclusions and the importance of the work. In studies with practical applications, the discussion includes the implications for society. For use in a 3MT we can combine the sections in this way: (i) introduction; (ii) methods and results; and (iii) discussion (including the implications for society). This leaves you with three main sections, corresponding to the 3 minutes of the presentation. Not coincidently these sections correspond well with the three acts in the three-act structure we discussed in the last chapter.

---

[1]https://threeminutethesis.uq.edu.au/resources/judging-criteria

Act 1: current state of knowledge = the problem you address (introduction)

Act 2: conflicting evidence; a new hypothesis is formed; data collection = what you did, and what you found (methods and results)

Act 3: conclusions = importance of your work and its implications for society (discussion)

In a 3-minute presentation you should spend approximately 1 minute on each act. We will jump right in by working through two examples.

 **Dr. Joshua Chu-Tan** (The John Curtin School of Medical Research, Australian National University), *Targeting the Root of Vision Loss*—2016 Asia-Pacific 3MT® Winner (Chu-Tan, 2016)

> "If I asked everyone here to think about the one sense that you couldn't live without, I'm willing to bet that most of you would have immediately thought of your sight."

One of the most important things you can do at the beginning of your talk is to engage your audience's imagination. Asking your audience to imagine something draws them into your presentation. From being a spectator, they become a participant. In some venues, speakers will ask the audience a question and wait for their response. A short back-and-forth will ensue. The 3MT format does not allow this, so Joshua engages the audience by asking and answering his own question.

> "That's because our vision, and what we see, plays such an integral role into how we perceive the world around us. Now, the tissue responsible for that is known as the retina."

These sentences provide a segue to the more technical part of the talk. He introduces the term retina and explains why it is relevant to our sense of sight. Though most audience members may already know the function of the retina, his explanation signals that he is going to define the technical terms, so they can relax and enjoy the presentation.

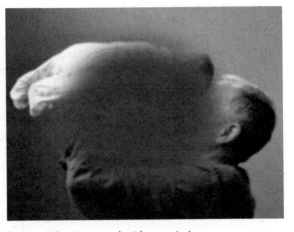

> "But what if I told you that a specific part of your retina was slowly damaged as you aged? What if this part of your retina was responsible for your color

©Joshua Chu-Tan, used with permission.

perception as well as your central vision, and that this damage leads to your visual field looking like this [points at slide]? Well, this is exactly what happens in age-related macular degeneration, or AMD."

Joshua again uses questions to engage the audience. This time he uses them to present the background detail that the audience needs to understand his research. He introduces the concept of the macula, the part of the retina that is damaged in AMD. Notice that he avoids the use of unnecessary technical terms. When he uses a technical term, he defines it.

"This disease is the leading cause of blindness in the developed world. This disease cost the Australian economy up to five billion dollars per year, and the most common form of this disease [1 minute] has no cure. AMD is a serious problem."

This is the 1-minute mark. In this last section he has explained the seriousness of the disease, its economic consequences, and the fact that it has no cure. What we do not hear in this written version is Joshua's dramatic vocal presentation. He uses frequent pauses, gestures, and changes in intonation to stress the seriousness of the problem. He slows dramatically down for the words "has no cure," looks directly at the audience, and gestures to emphasize his point. This is how Joshua shows his passion for his subject. He lets the audience see how much he cares. At the end of the first minute he has drawn the audience into his research, provided the background they need to understand his work, and explained why it matters. He has accomplished these things in a very vivid way.

"Now due to the many factors that are at play, therapies are very hard to come by. But I'm researching a potential gene therapy using tiny molecules called microRNAs that are like the gods of gene regulation and have the ability to bypass this problem."

At the beginning of the second minute he explains his research and introduces the term microRNA. To help the audience understand this term he refers to microRNAs as "the gods of gene regulation." Although not technically accurate, this simile tells the audience that microRNAs are powerful gene regulators. With this holistic understanding established, he moves on to explain microRNAs in more detail.

"How? Because a single microRNA can bind to multiple targets."

Saying that microRNAs are the gods of gene regulation does not explain how they work. It is a nice simile, but it is incomplete. Joshua must now go on and briefly explain how they work. Why are they like gods? They are like gods because they can bind to multiple targets. What does that mean?

"So instead of controlling just the one gene, we're controlling its entire pathway."

It means they are not just controlling a single gene they are controlling a whole pathway of genes. But again, what does that mean? What does it mean to control a whole pathway of genes?

"Instead of superficially cutting just the stem of the weed we're ripping out the whole root."

To cement the explanation, Joshua uses another simile. He compares the action of microRNAs to ripping out a weed by the roots. This is a vivid image that gives the audience a very clear metaphor for the action of microRNAs. They attack the root of the problem.

"So, in the case of AMD we know that the inflammatory pathway plays a major role. So we've injected an anti-inflammatory microRNA into the eye, and what we saw was a decrease in genes responsible for inflammation, cell death as well as a slowing in the damage progression of the retina. [2 minutes] What we've now done is identified multiple microRNAs that can each regulate pathways leading to AMD, and by injecting a cocktail of these molecules we can slow the progression of the disease, and hopefully halt vision loss."

This is 2 minutes and 17 seconds into the talk. At the end of 2 minutes he has finished presenting his methods and has presented the first, and most important part of his results. In the next 17 seconds he elaborates on his results and presents his conclusions. What remains for the last approximately 45 seconds is to explain the importance of these results.

"Now I want everyone to look at this image again [points at his slide]. Imagine living the rest of your life with vision like this."

To explain the importance of his results he returns to the beginning of his talk, reminds the audience of the severity of the disease, and once again points to his slide.

"So, the next time you're with your loved ones, study their faces, their every feature. Commit that to memory. Cherish that image, because one in seven of you will lose that ability if nothing is done about this disease."

He intensifies the audience's experience by making the story personal. He draws the audience in by asking them to imagine living their life with macular degeneration.

"With the use of microRNAs, my hope, and my goal, is that for millions of people this image can become clear."

He finishes his talk in just under 3 minutes by repeating the goal of his research.

Looking at the talk as a whole we see that it follows the classic hero's journey. The hero faces an intractable problem, develops a unique and innovative solution, and changes the world. He begins in the familiar world that we know, enters the woods, faces the monster, and returns with a boon for society. In the first minute he sets up the problem and emphasizes its severity. In the second, he describes his approach using metaphors and similes to explain the technical details. In the third, he explains what his solution means for society. The fit with classic story structure is remarkably accurate.

Let us look at another example.

 **Dr. Yasmin Mustapha Kamil** (Faculty of Engineering, Universiti Putra Malaysia), *Dengue Detective*—2018 Asia-Pacific and People's Choice 3MT® Winner (Mustapha Kamil, 2018)

"Have you ever been bitten by mosquitoes? Literally they suck; and when they bite, they make us itch."

Like Joshua, Yasmin begins with a question for the audience. In Joshua's case he used a question that, when answered, led the audience directly into his research. In this case the question has a yes or no answer. That means the audience does not have to imagine quite as much. Yasmin does most of the work for them. She tells them that mosquitoes suck, and that they make us itch. This is a good opening, but it would be even stronger if she had left more work for the audience. A slightly stronger opening might be: "What did it feel like the last time you were bitten by a mosquito? [pause] If you are like me, it hurt … then it itched for days." This formulation requires the audience to imagine the experience. It draws them deeper into the presentation. Both openings are good. Both are sensory. Yasmin will make use of this sensory experience when she comes back to it again at the end of her talk.

"But more than that, they transmit deadly diseases across the globe, including dengue. In a year 390 million people fall victim to dengue. That's like 16 times the population of Australia today; and 70% of deaths caused by the virus are due to one reason – a delay in detection."

Because of the title, the audience already knows that the talk will be on dengue. Yasmin now makes a connection between her opening remarks and this disease. She does that here and explains why the disease is so serious. She ends with a nice segue to introduce her research on the detection of the disease.

"I was a victim of dengue myself. Horrible experience. I had a high fever for three days and the doctors, like the mosquitoes, took my blood again and again; and it was not until the fourth day that they could finally confirm I had the infection and start proper treatment. But by then I was already too weak to even drink on my own. I had to be put on drips for a whole week. [1 minute] I felt helpless and afraid, but the worst part was having to witness other victims in my ward succumb to dengue just because they were not treated in time. I was lucky to survive, and I felt that nobody should die from something as trivial as a mosquito bite. Right? And so, I dedicated the next few years of my life to find a solution."

We are now 1 minute and 27 seconds into the presentation. In this last section Yasmin has nicely explained why she became interested in dengue and why her research is so important to her. Sharing these personal details makes a connection with the audience, but it also takes time. She is halfway through the talk and she has yet to explain her research. She has, however, done such a good job of explaining its significance that she does not have to spend much time on this at the end of her talk. The audience is primed to understand the importance of her work.

"What I have developed is a dengue sensor which is able to detect the virus more accurately and in a much shorter time. Meet my dengue detective [points to her slide]. It holds three basic

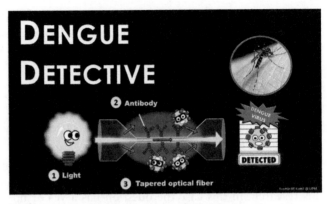

©Yasmin Mustapha Kamil, used with permission.

components: light, antibodies, and a tapered optical fiber, which has not been used before; and all it needs from a patient is one tiny drop of blood."

Although she uses the technical term antibodies, it is a term that she rightly assumes that most of her university-educated audience will understand. She does not need to define it , but she does need to explain how her dengue detective works. She does this in the next section.

"Now let me tell you how it works. Envision an underwater glass tunnel. You know the ones you find at aquatic exhibitions where you walk through to see sharks or stingrays swimming around you. Well now visualize this tapered optical fiber [2 minutes] as that glass tunnel immersed in a patient's blood sample. And on the surface of this fiber tunnel I immobilize antibodies to capture the virus. Next, I transmit light to travel through this fiber tunnel and indicate the presence and quantity of the virus. And *voilà*, dengue is detected and quantified."

The explanation of her research has taken almost exactly 1 minute. We are now at 2 minutes 22 seconds. Like Joshua, she uses a simile to explain her work. She likens her dengue detective to a shark tunnel, which she assumes will be familiar to her audience. Unlike Joshua, she depends more on her slide to explain how her approach works. Its operation is not completely clear from her verbal description.

"This dengue detective holds great promise. Now let me tell you why. First it is highly sensitive and reliable. Second it is affordable for all clinics to use. Lastly and most importantly it is able to reduce the detection time from four days to just 15 minutes, which gives dengue victims a greater chance to survive. This technology is a huge step forward in the future of dengue diagnostics."

With just over half a minute to go, Yasmin moves on to the importance of her work and its meaning for society. She is very specific about the advantages of her technique and how it will affect patient care. The specifics are important because they are concrete. The more concrete you can be, the more real and immediate your research will seem to the audience and the better they will remember it.

"Mosquitoes will still suck but this sensor will detect the virus in time. Case closed."

She finishes her talk at 2 minutes and 57 seconds by returning to her opening statement. She reminds the audience of the sensory experience she elicited and links it to the advantages of her research. Returning to the opening gives the audience a sense of closure.

Notice that neither of these authors had completed their research at the time of their presentation. This is common for competitors in the 3MT. As long as you are far enough along that you have some results, you can present a 3MT.

Best Practices

Now that we have seen two examples, we are in a better position to discuss best practices. To be memorable stories should be simple, surprising, concrete, emotional, and credible (Heath and Heath, 2007). As we mentioned in the previous chapter, all research is surprising. Anything that changes the world surprises us, and the purpose of research is to change the world. You may have thought that there was no cure for macular degeneration; surprise, there is. You thought that dengue was hard to detect; surprise, it is not. Surprise is one of the easiest things to achieve in a scientific talk. The three-act structure gives it to us with almost no effort.

Simple

Both speakers use a three-act structure to divide their talk into sections. Joshua's introduction is almost exactly 1 minute long. He uses a bit more time to present his methods and results (1 minute 15 seconds), and then explains the meaning and significance of his research in the last 45 seconds. Yasmin uses 1 minute and 22 seconds to introduce her research, another minute to explain what she has done, and about 30 seconds to explain its importance. The differences are slight but can be instructive. The longer time devoted to the importance of his research at the end of Joshua's talk allows him to make a better connection with the audience. He does this by asking them to imagine the faces of their loved ones. This is a reference back to the beginning of his talk when he

asked them to think about the one sense that they could not live without. The best presentations end with a reference to the beginning. If you begin by talking about the environmental effects of mercury, you should end by talking about mercury, but now informed by your research. You end on the same point with which you began and show how your research has changed what we know.

Yasmin also ends her talk with a reference to the beginning. She explains the importance of her research for society but does so in a more objective way than Joshua. She cites facts about the effectiveness of her technique and only references the beginning of her talk in the last full sentence, which lasts 5 seconds. It is a nice reference and closes out her presentation, but it is less effective at connecting with her audience. The connection would have been stronger if she had said something like this: "Remember what it feels like to be bitten by a mosquito? They still suck, but my sensor will detect the virus in time."

Concrete

Both speakers use concrete language to explain their technical terms. They try to avoid technical language altogether but when they must employ a technical term they use a simile to clarify its meaning. Joshua explains that microRNAs are the gods of gene regulation because they rip out the problem at the root. Yasmin likens her dengue detective to the type of shark tunnel found in major aquaria. These similes draw on the listeners' everyday experience to explain the technical details. They are concrete. Concrete language is visceral. It is easy to understand. It helps the audience form a mental picture. When used effectively, it both clarifies meaning and draws the audience into the presentation. Concrete language is a very effective tool for audience engagement.

Emotional

Research that your audience cares about, that elicits an emotional response, is more likely to be remembered than work that they find uninteresting. Let me give you an example. Some years ago I participated in a friendly 3MT-style people's choice competition at the Alan Alda Center for Communicating Science at Stony Brook University (Stony Brook, New York). One of the first speakers was an Australian physician who worked with neonatal infants. His slide was of a neonatal baby, intubated, covered with sensors, sleeping in an incubator. He won the competition as soon as this image appeared on the screen. If you work in a medical field you are lucky and you will seldom have to worry about eliciting an emotional response. However, for many of us we will have to find a way to help our audience care. We can do this by telling a personal story about our research, as Yasmin did. We can also use changes of intonation, expressive vocal patterns, gestures, and facial expressions to show the audience that we care. Joshua used this method. A third method is to use humor. Humor can be used effectively to get a response from your audience, but this method is fraught with difficulties. You want the audience to laugh because they are surprised. You do not want them to laugh at you or your research. To make humor work you must find a surprising way in which to make a point, a way that subverts the audience's expectations. It does not have to be, and probably should not be, laugh-out-loud funny. It does need to be surprising. My presentation in the aforementioned competition was on the use of images to replace written descriptions in certain scientific applications. Given this topic it would have been predictable if I had said "a picture is worth a thousand words." This would have met the audience's expectations, not subverted them. It would not have been funny. What I said was "There is a problem with descriptive terminology. [pause] What color is brunette [and I raised my arms with an exaggerated shrug of the shoulders]?" I then went on to give several technical examples. This simple example had members of the audience smiling. Because of this emotional response they were more attentive to the rest of my talk. One of them came up afterwards and told me it was a particularly good example.

brunette

Credible

Your audience will be more willing to believe you if they find you credible. There are several ways to establish this. If you are associated with a university like Harvard, Cambridge, or the University of Melbourne, you will have credibility just by showing up. Facebook was so successful partially because it started at Harvard and was initially restricted to Harvard students. This allure attracted others as the site began to expand. Personal experience with your subject can also establish credibility. You will get a credibility boost if you are a lung cancer survivor speaking about lung cancer research, especially if you share something of your personal experience. Yasmin gives a nice example of how to do this based on her experience with dengue. If your research has been successful in an area where many others have failed, you will also get a credibility boost. Joshua used this method when he emphasizes that macular degeneration "has no cure." It has no cure, but then he presents one.

Another way to establish credibility is to appeal to the direct experience of the audience. One way of doing this is to use questions to engage your audience's imagination. Questions quickly align the listener and speaker's beliefs about a topic. Both Joshua and Yasmin open their presentations with a question. Joshua asks: "If I asked everyone here to think about the one sense that you couldn't live without …" Yasmin asks the audience if they have ever been bitten by a mosquito. Both questions require the audience to be active. They ask the audience to transform an abstract idea into experience. When Joshua asks the audience to imagine the one sense that they cannot live without, the audience members perform a silent inventory of their senses. This transforms the abstract idea that your sight is your most valuable sense into a concrete, visceral experience. These types of experiences are much easier to remember. In Yasmin's talk the answer to her question is a single word but this word (yes) elicits the experience of being bitten by a mosquito. Without this imagined experience a simple yes/no question would be much less effective. The audience would quickly answer the question and revert to passivity.

If you are giving a talk where you can use props, you can increase your credibility by creating an experience for the audience. In the 1980s the Beyond War movement made the destructive power of the world's nuclear warheads tangible by dropping one tiny metal ball for each Hiroshima-equivalent warhead into a metal pail. The cacophony caused by the 5000+ metal balls made this number real in a way that statistics never could (Heath and Heath, 2007). All these methods allow you to make a personal connection with the audience, which makes your work more credible.

Image Use in a Three Minute Thesis

The 3MT allows the use of a single slide. Given this restriction, you must resist the temptation to clutter your slide with data. This will only confuse the audience and distract them from your talk. You want the audience to be focused on you, not your slide. A great slide will repeat the central message of your talk in a visual medium. This is what you should strive for. Let us look at some examples. We will start with Joshua and Yasmin's slides because they provide a nice contrast.

Joshua uses a single image without any text. The center of the picture is grayed out to simulate what happens in macular degeneration. When he points at the image, it is a powerful reminder of the problem that he is working to solve. The image succinctly summarizes his core message. It delivers this message in a different modality than his verbal presentation and therefore reinforces it.

Yasmin's slide uses a different approach. The three areas of her slide each deliver a different message. The text reminds the audience of her title. The picture of the mosquito references the first point in her presentation. At the bottom are three images that read from left to right and explain the operation of her dengue detective. This slide takes a good deal of thought to understand. Unfortunately, the audience will be trying to understand it while they listen to her speak. The situation would be different if there were less text on the slide and the images at the bottom could be made to appear in order. In that case the lightbulb would appear first, followed by the optical fiber and antibody, and then the detected icon. Unfortunately, the 3MT rules prohibit animations so the audience is left to decipher the slide on their own. Her verbal explanation helps, but a part of the listener's attention will still be spent trying to decipher the images when it should be directed to her explanation. Another approach might have been to use a single image of an engorged mosquito. This would reinforce the dangers of dengue and provide a strong link to Yasmin's opening where she asked the audience to imagine being bitten by a mosquito. Coupled with her

Photo credit James Gathany, public domain[1]

clear explanation of how her dengue detective works, an image like this would provide a visual and visceral reminder of the importance of her work.

Let us look at another example. The following two slides were created during one of the authors' workshops. Participants prepared a slide before the workshop and then revised it as we worked together. The first slide is a typical starting place for most students. It shows a sequence of images intended

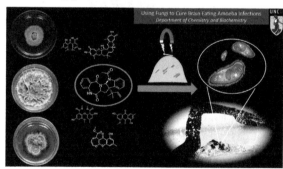

Kristof Cank, used with permission

to explain the student's research on the brain-eating amoeba (*Naegleria fowleri*). The second slide has no text at all (see next page). It shows a single image of a woman immersed in water up to her nose, the conduit by which the amoeba enters the body. Which slide better conveys the author's meaning? If the audience was only going to see a slide and not hear his presentation, the first slide would be much better. However, the author summarized everything on this slide in his talk. That means that the audience will be trying to listen to him and interpret the slide at the same time. This will interfere with their understanding. Given the fact that he will be presenting his work verbally, the second slide is much better. It says "There

[1]https://phil.cdc.gov/Details.aspx?pid=1969

Ryan Christodoulou on Unsplash[2]

is a problem when your nose is in open water. Watch out." This message is much more powerful than that delivered by the first slide. It is powerful because the message is delivered on two separate channels. It is delivered to our verbal-auditory brain by his verbal presentation and to our visual brain by the image. This is a very powerful way of delivering a message. It allows the audience to stay focused on your message.

Using text on a slide disrupts the listener's ability to understand your meaning (Morgan and Whitener, 2006). Look at these two examples of a turtle and decide which is clearer. As you imagine each slide appearing on the screen, imagine the speaker saying, "this is a turtle." Which image is more effective? Take a minute to write out your answer.

Photo credit NOAA, public domain.

Most people find the image without text to be more effective. This is because we process images and text with two different parts of our brain. These two areas do not easily talk to each other. Adding text to an image disrupts our visual processing of the image (Hantsch *et al.*, 2012). Our eyes are drawn back and forth between the text and the image so that we are never able to fully concentrate on either (Mayer and Moreno, 1998). Placing text on an image asks the audience to do two contradictory things: read the text and look at the image. Then, if you are speaking, the audience must also listen to you and so do three contradictory things: (i) read the text; (ii) process the image; and (iii) listen to what you are saying. It is impossible to do all three things at the same time. The most common response is to stop listening. This is the last thing you want to happen! You want to be the center of attention. You want your audience

---

[2]https://unsplash.com/@misterdoulou; License: https://unsplash.com/license

to listen to and understand what you are saying. An image with text on it interferes with this process.

Preparing a Three Minute Thesis

Here are some things to keep in mind as you prepare your talk.

1. Decide what you want your audience to know when they leave. They will remember one thing. What do you want that to be? In Joshua's case it was that he had a solution to the devastation of macular degeneration. In Yasmin's case it was her development of a new dengue detective.
2. Based on your answer to #1, create your slide. The slide should have a single image. There should be no text. The image should say the same thing, visually, as you will say in words. Free, high-quality images can be found at Wikimedia Commons, unsplash.com and pixabay.com.
3. Create your talk based on a three-act structure, keeping your central point in mind.
4. Be sure to use concrete language and to end your talk where you began.

Other Short Forms

The main differences between types of short talks are their length, whether visuals can be used, and if props are allowed. Lengths of the most common types of short talks range from 90 seconds (Perfect Pitch; Engineering Research Centers, 2021) to 3 minutes. The difference between a 3-minute and 90-second talk is one of emphasis. In a 90-second talk you will have less time to introduce your subject and explain its significance. The second act of your talk will also need to be shortened, but it should retain the substance of your message. You should spend between 15 and 20 seconds introducing your research, 60–70 seconds describing what you have done, and 5–15 seconds explaining its significance. You can spend less time explaining the significance of your work if you have done a good job introducing it at the beginning of your talk. Yasmin's introduction is quite long but it allows her to spend less time at the end discussing the significance of her work.

You must be very careful if you decide to use a video in your talk (Three Minute Wonder; Institute of Physics, 2021). It is very tempting to produce a flashy animation that will captivate your audience and distract them from your message. To avoid this temptation, you should remember that every second the audience is watching the video they are not listening to you. You should always be the center of attention. All the important information should come through you. This will help the audience make a personal connection with your work, which will increase their understanding. Without this connection your work can seem dry, academic, and unrelated to life. With it your work will seem interesting, relevant, and more related to their concerns.

It is probably better to avoid videos in very short talks. If you decide that one is essential, keep it short and related to your main message. Also think carefully about the image that the audience will see when the video is not playing. Most video editing software will allow you to choose a custom image, a poster frame, that is displayed before the video starts. This image should be chosen based on the same criteria we discussed above. It should be a visual summary of your main point. Check with your presentation rules to determine if you can use an image that is not part of the video. If the first and last images must be part of the video, you can put them there with video editing software like OBS (Open Broadcast System—available at: https://obsproject.com/). In this way the audience will always be looking at an image that is relevant to your presentation.

The same basic principles apply if you are presenting a talk that allows props but not images (e.g. FameLab (British Council, 2021) and Flash Talks (Appiah, 2020)). The best prop will be one that restates your central message in the same way as does the single image in a 3MT. Unfortunately, this is almost impossible to achieve. It is much easier to use a prop as a metaphor for some concept that is central to your presentation. For instance, Sauradeep Majumdar uses two differently colored balls to represent atoms of copper and zinc in his award-winning FameLab talk on using nanomaterials to capture CO_2 (Majumdar, 2020). Virtually no explanation is necessary for the audience to understand these props. Sauradeep only has to hold up the balls and say "these atoms" and he can go on with his explanation. The simplicity of his props allows him to focus on his meaning. He does not need to explain them because their meaning is intuitively clear. Contrast this with someone who holds up a complexly shaped object and identifies it as a protein. Unless the audience is composed of protein chemists, they will probably not know what a protein looks like. They will be confused by the prop. They will stare at it and try to figure out how it can represent a protein. This will distract them from your message. Your props must be simple, easy to understand, and related to your main point. They hinder your communication when they cause the audience to lose focus. If the audience has to work to understand your prop, it is too abstract. When you use simple props coupled with changes in vocal intonation, facial expression, and the use of exaggerated gestures, your audience will be surprised. When they are surprised, they will remember your talk. If your props are simple, concrete, and surprising, they will also be credible. They will convey your meaning directly to the audience. Test your props on a trial audience to check their meaning. If the audience finds them difficult to understand, find simpler ones.

Appiah, B. (2020) How to give a science flash talk. Script Practical Guide. Script. Available at: https://scripttraining.net/script-practical-guide/media-skills-for-scientists/how-to-give-a-science-flash-talk/

British Council (2021) FameLab: Science Communication Competition. Available at: https://www.britishcouncil.org/education/he-science/famelab

Chu-Tan, J. (2016) Targeting the Root of Vision Loss. 2016 Winner Asia-Pacific 3MT Final. The University of Queensland, Brisbane, Queensland, Australia. Available at: https://threeminutethesis.uq.edu.au/asia-pacific-past-finalists

Engineering Research Centers (2021) Perfect Pitch Guidelines. National Science Foundation (NSF) Engineering Research Centers, USA. Available at: https://erc-assoc.org/content/perfect-pitch-guidelines

Hantsch, A., Jescheniak, J.D. and Mädebach, A. (2012) Naming and categorizing objects: task differences modulate the polarity of semantic effects in the picture–word interference paradigm. *Memory & Cognition* 40, 760–768. doi: https://doi.org/10.3758/s13421-012-0184-6

Heath, C. and Heath, D. (2007) *Made to Stick*. Random House, New York.

Institute of Physics (2021) Three Minute Wonder Science Communication Competition. Institute of Physics (IOP), UK. Available at: https://www.iop.org/3mw.

Majumdar, S. (2020) How carbon capture can help tackle climate change. FameLab International Final 2020, Global Winner. *FameLab International*. Available at: https://youtu.be/wlc3STkKOyg?t=2352

Mayer, R.E. and Moreno, R. (1998) A split-attention effect in multimedia learning: evidence for dual processing systems in working memory. *Journal of Educational Psychology* 90, 312–320. doi: https://doi.org/10.1037/0022-0663.90.2.312

Morgan, S. and Whitener, B. (2006) *Speaking about Science: a Manual for Creating Clear Presentations*. Cambridge University Press, Cambridge, UK.

Mustapha Kamil, Y. (2018) Dengue Detective. 2018 Winner & People's Choice Asia-Pacific Final. The University of Queensland, Brisbane, Queensland, Australia. Available at: https://threeminutethesis.uq.edu.au/asia-pacific-past-finalists

The University of Queensland (2008-present) 3MT® Three Minute Thesis. The University of Queensland, Brisbane, Queensland, Australia. Available at: https://threeminutethesis.uq.edu.au/

Elevator Pitches

4

An elevator pitch is a short, 15–90 second, polished summary of your work. The idea is to present your research in the time it takes to ride an elevator.

You will need to explain your work to different groups, so you will need several versions of your pitch. Each will be tailored to a particular audience. You might use one pitch when building collaborations or networking. Another might be a persuasive pitch that you use to attract funding or when talking to a policy maker. A slightly different version can be used when talking to friends or family about your research. Finally, you might need a version to present your research in an elevator pitch competition or in a short online video. You can even create a version for use during a job interview. The thing that all these uses have in common is that they communicate your message very succinctly, in a minimal amount of time.

We will divide our treatment of elevator pitches into two sections. The first deals with informal pitches such as you might use in private conversations. In these settings, your most common goal will be to create a favorable impression and open the door to further conversation. We will then go on to discuss more formal pitches. These pitches might be used in an elevator pitch competition or an online video. In formal pitches the focus will be on your research. Your goal will be to present your research clearly and concisely. When you do, you will leave the audience with a favorable impression of you and your abilities. The difference between these two types of pitches may seem slight, but there are major implications for how you design your presentation. In the first case you want your focus to be on the other person. Your pitch will be more personal and will be delivered one-on-one or to a very small audience. In the second, your focus will be on your research. Be sure you know which goal you are working towards as you develop your pitch.

The next step in preparing your pitch is to identify your audience. Are you speaking to a layperson who knows little about scientific research? Are you speaking to a scientist who works in a different field? Are you speaking to someone who knows a great deal about your research area? You will design your pitch differently for each of these audiences. You need to eliminate all jargon and use metaphors and similes when speaking to a layperson. Scientists who work in a different area can be expected to be familiar with some technical terminology but will need disciplinary jargon explained. For instance, a physicist might be familiar with the basic concepts of genetics but will be lost the minute you start talking about *Antp* (the Antennapedia gene). If you are speaking to someone in your discipline, they will not only know all the jargon but will expect you to use it intelligently. This is particularly important when you are introducing yourself during a job interview. These introductions must be both short and technically accurate.

Informal Pitches

Let us begin with a simple informal pitch. We will start with a brief statement that you might use to introduce yourself at a professional networking event outside your discipline. Your goal here is to open the door to further conversation. Although we will use the work of the three scientists we introduced in earlier chapters, we wrote these pitches and they have not been approved by the scientists.

> Hi, I am Pieter Visscher. I study the oldest forms of life on Earth.

> Hello, I am Irina Vetter. I study neurotoxins to find new ways of controlling pain.

> I'm David Nash. I work on the origins of Stonehenge.

Notice how these sentences focus on the most important aspect of each scientist's research. In the theater we would call this the main character (Olson, 2020). Dr. Vetter's sentence is particularly instructive. She introduces herself as working on pain not on *Dendrocnide* stings. Her work with *Dendrocnide* stings is a means to an end. She is interested in these plants because of what they tell us about pain. Therefore, her opening sentence is about pain not about the plants. Her focus on pain has the added benefit that it speaks to a general audience. Few people have heard of the Australian stinging tree, but everyone has experienced pain. Her statement will capture her listener's attention and make it easy to continue the conversation. The statements by Dr. Visscher and Professor Nash work in the same way. Their one sentence pitches are designed to generate questions. What are the oldest forms of life on Earth? Who built Stonehenge? The resulting questions will open the door for further conversation.

Now, use the following template to introduce yourself.

Hello, I am (your name) . I study/work on (the most important aspect of your work) .

You should introduce your research in the most general way possible. For instance, if you work on the *CG4050* gene in *Drosophila*, introduce this by talking about how your work relates to human noc-

turnal seizures (Oriel and Lasko, 2018). You might say "I am a geneticist. I work on genes linked to nocturnal seizures and intellectual disability." This is sure to generate interest and lead to further conversation. Once you are in conversation you can explain that you work on the fruit fly version (ortholog) of a gene involved in these traits. You can explain why fruit flies serve as a good model for human health. If you are speaking to someone during a job interview or to someone in your field, you can use technical terms: "I work on the *Drosophila* orthologs of genes associated with the genetic disease periventricular nodular heterotopia, which produces nocturnal seizures." You will need to go into more detail about periventricular nodular heterotopia, but this statement will establish your credibility as a scientist.

In addition to networking events, you can use a short pitch to introduce yourself before you ask a question at a scientific talk. As always, be sure you tailor your introduction to your intended audience. Use a general introduction for a talk outside of your field. Reserve your technical introduction for talks in your specialty. This will give you valuable exposure and perhaps open the door to conversations after the talk.

Let us now assume that you have introduced yourself and that you have been asked a follow-up question. You have already given a high-level overview of your work. Now you need to fill in some details. You should begin with the core message that you want your conversation partner to take away. Your goal is to focus on the most important point you want to make. In Chapter 6, on posters, we will work on writing one sentence summaries of your research. You want to continue your pitch with a similar summary. This gives your listeners the information they need to understand your research if the conversation continues. Here are some examples from our three scientists.

> **Visscher:** I study the oldest forms of life to learn how they survived without oxygen. The mats of bacteria that existed on the early Earth were incredibly stable. They lasted for such a long time that they built rock formations called stromatolites. Almost all existing organisms depend on oxygen, but for at least 1 billion years there was no oxygen in the atmosphere. I want to know how bacteria captured energy in a world without oxygen.

> **Vetter:** People who suffer from chronic pain have poorer health and a poorer quality of life. But we still have a lot to learn about how our nervous system senses pain. I study neurotoxins, especially those derived from plants, to better understand how we perceive pain in the hope of finding new ways to control it.

> **Nash:** Stonehenge is a national treasure. The smaller "bluestones" have been traced to localities in Wales. It has long been suspected that the large uprights and lintels originated from the Marlborough Downs, 25 km north of Stonehenge, but the providence of these larger stones is not known for certain. I tested the hypothesis that they originated from the Marlborough Downs to see if it is correct.

All these pitches extend the narrative and provide additional information. None are complete. They leave the door open for further conversation. It is not yet clear what Dr. Visscher means when he says, "I want to know how bacteria captured energy in a world without oxygen." He is interested in what the stromatolites used for an electron source in place of oxygen, but this has not yet been made clear. Dr. Vetter's statement is also incomplete. It has introduced the idea of plant neurotoxins but has not explained which plants and which neurotoxins are her most recent subjects. Professor Nash's statement is the most complete. It explains the specific hypothesis that he wants to test. However, it does not go into any detail about how he intends to do this. Nor does it reveal his results. All these presentations are fine at this stage of the conversation. Each scientist can continue their explanation if their partner is interested. If not, they have succeeded in their main goal of introducing themselves and their research.

Notice that all these elaborations all start with a clear reference to the main research subject. Visscher begins by speaking about the oldest forms of life on Earth. Vetter speaks of chronic pain. Nash speaks about the providence of the stones. The listener receives an overview of each scientist's research in the opening sentences. The details will come later. Also notice that each of these pitches follows a classic And-But-Therefore (ABT) structure. Go back and read each one and mark the AND, BUT and THEREFORE transitions. Our answers are in the box below.

Answer to exercise

Visscher: I study the oldest forms of life to learn how they survived without oxygen. (AND) The mats of bacteria that existed on the early Earth were incredibly stable. (AND) They lasted for such a long time that they built rock formations called stromatolites. Almost all existing organisms depend on oxygen, (BUT) but for at least 1 billion years there was no oxygen in the atmosphere. (THEREFORE) I want to know how bacteria captured energy in a world without oxygen.

Vetter: People who suffer from chronic pain have poorer health (AND) and a poorer quality of life. (BUT) But we still have a lot to learn about how our nervous system senses pain. (THEREFORE) I study neurotoxins, especially those derived from plants, to better understand how we perceive pain in the hope of finding new ways to control it.

Nash: Stonehenge is a national treasure. (AND) The smaller "bluestones" have been traced to localities in Wales. (AND) It has long been suspected that the large uprights and lintels originated from the Marlborough Downs, 25 km north of Stonehenge, (BUT) but the providence of these larger stones is not known for certain. (THERFEFORE) I tested the hypothesis that they originated from the Marlborough Downs to see if it is correct.

Your turn

Write a short pitch describing your research to a general audience. Start with a restatement of what you study and why you study it. Explain why your subject is important. Add some background information (some ANDs). Explain why there is a problem with our current knowledge (the BUT), and end with what you have done, or hope to do (your THEREFORE). If you like, you can begin by writing your pitch to someone in your field. Then go back and take out the jargon and explain what you do in non-technical terms. Aim for no more than 30 seconds. Put down the book and try this now. We will be here when you get back.

What methods and results to include in your pitch

None of the pitches have so far contained any details about methods or results. There are several reasons for this. First, if you are a student you will probably not yet have results. We wanted to show that you can still pitch your research even if you are still in the process of data collection. Second, in most networking events you are unlikely to have the opportunity to go into detail about your methods or results. If you can, it will probably be in response to a question. If someone asks a question you can explain more about your work. We cover more about how to present your methods and results in the next section on formal pitches. Once you have prepared your formal pitch, you can use some of the same material in informal settings.

Formal Pitches

Formal pitches are more appropriate for use in a competition, in an online video, or as a persuasive pitch that you use to attract funding or when speaking to a policy maker.

In more formal situations there is no opportunity for back and forth with your audience. You have the stage to yourself and must hold their attention. The longer versions of your informal pitch can serve as the opening for these more formal presentations.

Formal pitches are delivered without slides and are very short. They are like short versions of the 3-minute talks we covered in the last chapter. The most common length is 60 seconds, or less. For these pitches you will not need to introduce yourself. This will already have been done by the organizer. This allows your presentation to focus on your research. Your goal is still to get your audience to remember you, but now you must accomplish this solely through the clarity of your presentation.

We will continue our use of the And-But-Therefore structure to create formal pitches. You can think of these pitches in this way: There is this situation AND these facts pertain to it, BUT there is also this problem. THEREFORE I am working on this new approach that has the power to address the problem. If more details are needed, you can add them with IF/THEN phrases (Olson, 2020). These additions are particularly helpful if you work in a complex field that is unfamiliar to your audience. Here are three examples based on Dr. Visscher's work. Each one is directed at a different audience. Notice how the use of technical terms increases as we move from a general to a disciplinary audience.

General public: Can you imagine what the world was like before oxygen? All the plants and animals with which we are so familiar depend on oxygen. They could not and did not exist in that world. During this period only large mats of bacteria were present living in shallow pools. These bacterial mats used dissolved chemicals to help them capture energy from the sun. Even after years of study we still do not know for certain which chemicals they used. If we could find similar bacterial mats alive today, we could study them to gain insight into the early Earth. I searched for these mats in the most extreme environments on Earth. Eventually I found them in the high Andes. Study of their chemistry revealed that they use arsenic in the same way that modern plants use oxygen. This is important because arsenic is not used up in this process. It is cycled between two forms. This means that it can be reused for millions of years. Because of this the bacterial mats could be stable for the billion years before oxygen was present in the atmosphere.

Professional audience: The Earth existed for at least a billion years before there was any oxygen in the atmosphere. During this time life consisted of mats of bacteria growing in shallow pools. These bacteria captured energy from the sun using dissolved inorganic compounds as part of their raw materials. However, we do not know which compounds they used. If we knew this, we would have a much better understanding of how these bacterial mats were able to persist for millions of years. To solve this problem, I searched for an existing environment that was as much like the early Earth as possible. My colleagues and I found it in the Atacama Desert in the high Andes. NASA used this area to test its Mars rover because it is the closest place on Earth to Mars. We found that the bacteria growing in the shallow pools in the Atacama Desert cycle arsenic in the same way that existing plants cycle oxygen. Water is consumed in photosynthesis, producing oxygen. Oxygen is consumed in respiration, producing water. The bacteria in the Atacama Desert cycle two forms of arsenic in a similar way. One is produced in photosynthesis and then cycled back to its original form in respiration. The fact that arsenic is cycled and not consumed means that these bacteria can continue to grow indefinitely, even over millions of years.

Disciplinary audience: For at least a billion years before there was oxygen, life consisted of mats of photosynthetic, carbonate secreting bacteria growing in shallow pools. The rocks left behind by these mats are called stromatolites. They are the proof that there was life on the early Earth. Modern plants use water as an electron source in photosynthesis and produce oxygen as a byproduct. Then, in respiration, they use oxygen as an electron acceptor and produce water. We do not know what compounds the ancient bacteria used for these processes. If we knew that, we would have a much better idea of how the stromatolites were able to persist for millions of years. To solve this problem, I searched for an existing environment that was as much like the early Earth as possible. My colleagues and I found it in the Atacama Desert in the high Andes. We found that the bacteria growing there in shallow pools use arsenite as their source of electrons. They harvest electrons from arsenite to form arsenate. The arsenate is then cycled back to arsenite in respiration. The fact that arsenic is cycled means that the bacteria can continue to grow indefinitely, which allows the stromatolites to persist for millions of years.

Your turn

Now that you have seen three examples, go back and mark the transitions in the pitch to the professional audience. Mark the AND, BUT and

THEREFORE sections and make a note of where an IF/THEN statement is used. Our answer is given in the box below.

Answer to exercise

Professional audience: The Earth existed for at least a billion years before there was any oxygen in the atmosphere. (AND) During this time life consisted of mats of bacteria growing in shallow pools. (AND) These bacteria captured energy from the sun using dissolved inorganic compounds as part of their raw materials. (BUT) However, we do not know which compounds they used. (IF) If we knew this, (THEN) we would have a much better understanding of how these bacterial mats were able to persist for millions of years. (THEREFORE) To solve this problem I searched for an existing environment that was as much like the early Earth as possible. My colleagues and I found it in the Atacama Desert in the high Andes. NASA used this area to test its Mars rover because it is the closest place on Earth to Mars. We found that the bacteria growing in the shallow pools in the Atacama Desert cycle arsenic in the same way that existing plants cycle oxygen. Water is consumed in photosynthesis, producing oxygen. Oxygen is consumed in respiration, producing water. The bacteria in the Atacama Desert cycle two forms of arsenic in a similar way. One is produced in photosynthesis and then cycled back to its original form in respiration. The fact that arsenic is cycled and not consumed means that these bacteria can continue to grow indefinitely, even over millions of years.

For more practice, go back to Chapter 2 and review Dr. Vetter's work on *Dendrocnide* stings. Then write a 1-minute pitch to the general public based on her work. Be sure to include AND-BUT-THEREFORE transitions and an IF/THEN statement. Our version is at the end of the chapter.

Outline of a Formal Pitch

Now that we have seen several examples, we can take a step back and look at some of the basic principles for use in creating a formal pitch. At the beginning of the pitch, you need to gain the attention of your audience. This is very important when speaking to the public. A technical audience, especially one in your discipline, will more naturally be interested in your work. You cannot assume this for the public. You must draw them in. One of the most effective ways of doing this is to ask a question (Boisvenue, 2013). Questions draw the audience out of their private thoughts and help them focus on your presentation. Questions call for an answer. Even though they cannot be answered verbally, your audience will answer the question in their heads.

Following your opening, use "AND" statements to provide additional background. Keep these statements short. Only include the information that your audience must have. Your "BUT" statement introduces your problem. It tells the audience that the current state of knowledge is incomplete. The more clearly you state the problem, the easier it will be for you to explain your results. Once you have stated the problem, you can use IF/THEN statements to add detail and further explain the importance of your work. These statements help your audience understand why they should care.

The presentation of your results begins with your "THEREFORE" statement and continues with your explanation of how your solution solves the problem. If you have completed your research, present your findings. If your work is in progress, tell the audience what you hope to find. It is better to avoid presenting expected results. It is much more powerful to say "I hope" than "I expect." Hope is a more universal emotion than expectation. Hope will draw in your audience. Recall Joshua Chu-Tan's hope for his research (Chapter 3). His hope for a cure created a great connection with his audience.

Persuasive Pitches

Persuasive pitches are similar to formal pitches but they end with some type of request (called "The Ask"). They also often allow for interaction with your listeners. They can be the starting point of a conversation. We will leave it to you to identify your most important need and add a brief statement to the end of your formal pitch. This and a slight change in emphasis in the final sentences are what are needed to transform your formal pitch into a persuasive pitch.

Some Final Tips

Dr. Bernadine Healy, the first female director of the US National Institutes of Health had some of the best writing advice of all time. A sign on her desk read "Strong verbs. Short sentences" (Gladwell, 2018). There is no better advice as you prepare your pitch. Keep your sentences short and use strong verbs. A pitch is not the place to say "perhaps" or "should." Write like you are the expert that you are. The fewer words you use, the better.

Using strong verbs and short sentences does not mean that your first draft will be perfect. We doubt a single sentence from the first draft of this book has survived unscathed. Our second advice is to write fast and revise extensively. Beginning is the hardest part of writing. It is much more important to write something, anything, then to write perfectly, or even well. If you have something, you can make it better. If you have nothing, there is nothing to improve. For instance, to write the pitch of Dr. Vetter's work in the box below we started with the description of her work from an earlier chapter and revised it. Our edits were so extensive that there is now very little relation between the two versions. We cut all the technical language and many of the details. After we were done, the pitch was still too long and so we went back and cut 20% more. Writing short pitches for the general public is hard. It is much easier if you have something to start with. Write something, even something technical, then revise it. Your work will go much smoother.

Answer to Exercise

One-minute pitch to the public based on Dr. Vetter's work on *Dendrocnide* stings

Did you know that twenty percent of the population suffers from chronic pain? My colleagues and I study poisons that affect the nervous system in the hope that we can find new ways to control pain. (AND) Given this interest, it was natural for me to work on the Australian stinging tree. (AND) Stings from this tree produce excruciating pain that lasts for hours. (AND) Painful flares can continue for days, and sometimes weeks. (BUT) However, compounds from related plants have been shown to cause much less severe pain. (IF) If we can identify the compounds that cause such extreme pain, (THEN) we may be able to find new ways to control chronic pain. (THEREFORE) To find the real culprit, my team and I isolated and tested potential toxins from stinging tree hairs. A compound that caused similar extreme pain turned out to be a very stable small protein. To be sure that we had the right compound, we created a synthetic version of the protein and found that it can reproduce the symptoms of a sting. But how was it doing this? Using isolated neurons, we showed that the protein specifically affected these cells. Other tests allowed us to confirm that the proteins were neurotoxins. When we went back and investigated where these proteins occur in the stinging tree, we found them in the hairs. This gave us confidence that we had found the correct compound. The proteins we found are similar to the neurotoxins found in spider and cone snail venoms, some of the most toxic of all venoms.

You may have noticed that there is a second BUT in this summary. The second BUT begins "But how was it doing this?" This is followed by a second THEREFORE that begins "Using isolated neurons." What is happening here? We have, in essence, created a two-act play. Each act has its own And-But-Therefore structure. This is the first clue about how to construct a longer presentation using a variation on the five-act structure. We will return to this in the next chapter.

Boisvenue, G. (2013) The art of the elevator pitch: a qualitative study on the key rhetorical features of a successful venture capital pitch. MSc thesis, Purdue University, West Lafayette, Indiana, USA. Available at: https://www.proquest.com/docview/1433828502

Gladwell, M. (2018) *Strong Verbs. Short Sentences.* Revisionist History: Season 3, episode 9. Pushkin Industries. Available at: https://www.pushkin.fm/episode/strong-verbs-short-sentences/

Olson, R. (2020) *The Narrative Gym.* Prairie Starfish Press, Los Angeles, California.

Oriel, C. and Lasko, P. (2018) Recent developments in using *Drosophila* as a model for human genetic disease. *International Journal of Molecular Sciences* 19: 2041. doi: https://doi.org/10.3390/ijms19072041

Applications to Longer Forms 5

Longer presentations can be designed based on the same principles we have learned in previous chapters. The five-act form works well for presentations of 15–20 minutes. Effective presentations of an hour or more can be constructed by adding additional acts. We will begin this process with a review of the five-act structure. From there we will move on to longer forms.

As we explore longer forms it is important to remember that our suggestions are meant as guidelines. We present them because they embody timeless principles that are widely applicable. Use the ones that work for you.

Perhaps the most important principle we have learned is the one embodied in the And-But-Therefore (ABT) structure. This principle emphasizes the major narrative transitions. The first is the transition between the current state of knowledge and the conflicting evidence (BUT). The second is the formation of the new hypothesis (THEREFORE). An example will demonstrate the importance of the first of these transitions.

Which of these two sentences is more engaging?

I did this AND this AND this, AND then this AND this, AND finally concluded this.

I did this AND this, BUT I had this problem AND so I did this AND finally concluded this.

The addition of even one BUT makes a narrative much more compelling. The repetition of ANDs lulls us to sleep. The BUT introduces contradiction. It engages our brain. We want  to know what the problem is and how to solve it. The BUT introduces narrative tension. It causes us to look for meaning. In combination with the five-act structure it helps create engagement. It helps us transform an endless list of facts into a story. It helps our work to have impact.

Let us take a deeper look at the five-act structure and apply what we have learned about the importance of transitions.

Five-act Structure for Medium-length Presentations

The flow of the five-act structure moves from the current state of knowledge, through conflicting evidence, the formation of a hypothesis, to data collection and conclusions. The outline of the structure looks like this:

Act 1: current state of knowledge
Act 2: conflicting evidence; new hypothesis
Act 3: first presentation of results
Act 4: problems and their solution
Act 5: conclusions

To make this form more interesting we can introduce at least one BUT in each of Acts 3 and 4. Act 2 already has a BUT in the form of conflicting evidence. To hold the audience's interest it is best to introduce this same tension into the other acts. Without it the acts can easily become a list of facts (ANDs). Each BUT serves to wake up the audience. The BUTs tell them that something important is about to happen. The revised structure looks like this.

Act 1: current state of knowledge
Act 2: conflicting evidence (BUT); new hypothesis
Act 3: results, BUT, results
Act 4: BUT, results, BUT results
Act 5: conclusions

Act 3 begins with the presentation of the first part of your data, perhaps with the results of your first experiment. The BUT in the middle of this act might be one of the problems that remained after you collected this data. Perhaps the data needed to be verified. Perhaps it was incomplete. Perhaps some part of it was unexpected and led to other questions. All of these are potential BUTs: but we still needed to verify these results … ; but we still needed to show … ; but this was not what we expected and so we … .The contradictions punctuate the act so that it becomes more of a story and less an endless presentation of data with no underlying narrative.

The break between Acts 3 and 4 should be something important. There should be some change in the narrative at the beginning of Act 4. This shift is the reason for a new act. A great way to begin Act 4 is with a question. Questions are powerful BUTs. They show that there is still something unknown. They hint at the possibility of failure. They tell the audience that something important is coming. We will see an example of this use of questions in our summary of Dr. Vetter's work, which we present next.

Neurotoxic peptides from the giant Australian stinging tree

Let us put Dr. Irina Vetter's study of *Dendrocnide* stings into a five-act structure with a few BUTs added for interest. We have broken up Acts 3 and 4 with several questions. Minor BUTs occur within each act. Major ones start a new act. To refresh your memory, we have added a few details to our summary of her work and reproduced it below. You might also find it helpful to review the three-act summary of her work in Chapter 2.

Dr. Vetter's interest in the molecular pharmacology of pain led her to work on the stinging trees of the Australasian genus *Dendrocnide* (Urticaceae). The compounds responsible for the painful and persistent stings caused by these plants reside in the epidermal trichomes that cover the plant surface. The pain lasts for several hours but painful flares can continue for weeks. Based on what she knew about the pharmacologically active compound in stinging nettle trichomes, Dr. Vetter assumed that this compound, a small molecular neuropeptide called moroidin, would also be found in the trichomes of *Dendrocnide* spp. However, when her team tried to isolate this compound from *Dendrocnide excelsa*, they found that it was not present. Also, when moroidin had previously been injected alone, it failed to produce the substantial long-lasting pain that occurs with contact with the plant. Although moroidin was eventually isolated from a different species (*Dendrocnide moroides*), injection of trichome extract purified to exclude moroidin still produced the symptoms of a *Dendrocnide* sting. These results, taken together, suggested that there were unidentified pharmacologically active compounds in *Dendrocnide* trichomes.

Based on these results, Dr. Vetter began work with Dr. Thomas Durek, Senior Research Officer and a chemist at the University of Queensland, and assembled a research team to look for these compounds. They isolated and purified the molecular constituents from *D. excelsa* trichomes and assessed the sensory effects of the resulting fractions. A late-eluting fraction of the extract elicited similar responses to a crude trichome extract. Mass spectrometry of this fraction showed the presence of miniproteins stabilized by disulfide bonds. These results suggested a possible causative agent, but verification was still required. Experiments with synthetic versions of these peptides showed that they can reproduce the symptoms of a *Dendrocnide* sting in a mouse model. The severity of the reactions suggested that the peptides might be neurotoxins. To investigate this possibility the team used mouse dorsal root ganglion neurons and fluorescence Ca^{2+} microscopy to investigate Ca^{2+} influx in cells. Consistent with their expectations, Ca^{2+} influx occurred in neurons but not in non-excitable cells exposed to the peptides. Patch-clamp electrophysiology confirmed that

the peptides work directly on voltage-gated sodium (Na_V) channels, indicating that they are neurotoxins. Matrix-assisted laser desorption/ionization imaging mass spectrometry (MALDI-IMS) demonstrated that molecules of the proper molecular mass localize specifically to the trichomes. Though this class of neurotoxins was previously unknown from plants, they are similar in structure to neurotoxins found in spider and cone snail venoms (Gilding *et al.*, 2020).

Five-act structure

Act 1: current state of knowledge
- a. Contact with *Dendrocnide* plants causes excruciating long-lasting pain.
- b. The causative agent may be moroidin.

Act 2: conflicting evidence; new hypothesis
- a. BUT: *D. excelsa* trichomes lack moroidin; purified extracts of moroidin do not produce the same long-lasting effects as the stings; injection of trichome extract purified to exclude moroidin still produced the symptoms of a sting (conflicting evidence).
- b. Other pharmacologically active compounds are responsible for the stings (new hypothesis).

Act 3: first presentation of results
- a. A late-eluting fraction of trichome extracts from *D. excelsa* elicits similar responses as the crude trichome extract.
- b. Mass spectrometry of this fraction shows the presence of miniproteins stabilized by disulfide bonds.
- c. BUT: Are these peptides responsible for the stings?
- d. Synthetic versions of the peptides are synthesized.
- e. Animal experiments with the synthetic versions reproduce the symptoms of a *Dendrocnide* sting.

Act 4: problems and their solution
- a. BUT: How do the miniproteins function to cause the stings?
- b. Fluorescence Ca^{2+} microscopy shows that Ca^{2+} influx occurs in neurons but not in non-excitable cells exposed to the peptides.
- c. Patch-clamp electrophysiology confirms that the peptides work directly on Na_v channels.
- d. BUT: Can we be sure that the miniproteins occur in the trichomes?
- e. MALDI-IMS demonstrates that compounds of the proper molecular weight localize specifically to the trichomes.

Act 5: conclusions
- a. The causative agents of *Dendrocnide* stings are neurotoxins similar to those found in spider and cone snail venom.

The result is clear, compelling, and will hold the attention of an audience.

Let us see how we would apply this form to Professor Nash's Stonehenge study. The five-act form is not quite as applicable to this study because there is no conflicting evidence. The hypothesis Professor Nash addressed was formulated long ago. The problem was that there was no way to test it. We begin with a summary of his work with a bit of detail added, then place it into a five-act structure.

Uncovering the origins of the sandstone megaliths at Stonehenge

The sandstone megaliths that form the massive uprights and lintels of Stonehenge have traditionally been thought to originate from the Marlborough Downs. However, this hypothesis has never been scientifically tested. Determination of the providence of the stones has been difficult because of the inability to destructively sample from the site. This changed in 2018 when a drill core from one of the megaliths (Stone 58) was returned to English Heritage. The return of this core allowed Professor Nash's team,

with the permission of English Heritage, to destructively sample a piece of stone from the unweathered center of this megalith. A few years previously Nash *et al.* (2013) had developed inductively coupled plasma mass spectrometry for use as a provenancing tool on silcrete rocks, the major component of the sandstone megaliths. The existence of this technique was a necessary condition for carrying out the research at Stonehenge. Before applying this technique, the team used non-destructive X-ray fluorescence spectrometry to investigate the chemical composition of the 52 remaining megaliths. The collection of this X-ray data was very challenging. The team was only allowed access to the monument when it was closed to the public. This meant several nights of data collection in the dark with headtorches in the freezing cold. The sheer size of the stones was also a challenge. The team needed to use a mobile scaffold to access the lintel stones. This required various health and safety assessments, and an archaeological assessment to ensure that there was no risk of damage to either the stones or the soil at the monument. Linear discriminant analysis of the X-ray data coupled with Bayesian principal component plots demonstrated that 50 of the 52 megaliths, including Stone 58, shared a consistent geochemistry and likely originated from a common source. When they were finally given permission to sample the returned core, they were able to proceed with their analysis. From a 7-cm-long section of this core they had to obtain enough material to make multiple geological thin sections as well as take samples for chemical and isotope analysis. They achieved this using precision rock-cutting equipment available from their partner universities. The team then used inductively coupled plasma mass spectrometry to destructively sample a small portion of the core. They compared the resulting geochemical signature with equivalent data for similar stones from across southern Britain. From this, they identified an area 25 km north of Stonehenge as the most probable source for most of the megaliths at the monument.

Five-act structure

Act 1: current state of knowledge
    a. The sandstone megaliths of Stonehenge are thought to have originated from the Marlborough Downs, but there has been no scientific test of this hypothesis.

Act 2: conflicting evidence; new hypothesis
    a. A meter-long drill core from Stone 58 is returned to English Heritage (conflicting evidence = data collection may now be possible).
    b. A new technique, inductively coupled plasma mass spectrometry, for provenancing silcrete rocks is available (new hypothesis = it is possible to providence the stones).

Act 3: first presentation of results
    a. X-ray fluorescence spectrometry data is collected.
    b. BUT: Data collection must be completed in the dark, on mobile scaffolding, in the freezing cold.
    c. The results show that 50 of the 52 remaining megaliths came from a single source area.

Act 4: problems and their solution
    a. BUT: The team still needs permission to destructively sample from the drill core from Stone 58.
    b. English Heritage grants permission to sample the core.
    c. BUT: Only 7 cm of the meter-long core is allowed to be sampled.
    d. Precision rock cutting equipment is needed to prepare the sample for analysis.

Act 5: conclusions
    a. Stone 58, and by correlation 49 other megaliths, is localized to an area on the south-eastern edge of the Marlborough Downs

We see that some accommodations had to be made to apply the form to this research. There is no conflicting evidence or new hypothesis. The return of the drill core and the development of the technique for provenancing silcrete rocks are the events that impel the research. Without these events the work would have been impossible. There is also no question or conflicting evidence in either Act 3 or 4. To create tension we have added information about the difficulties the team had to face to the middle of Act 3. The addition of this information, supplemented with a few well-chosen photographs, will provide a point of interest for the audience. It adds a human element to the talk. This is similar to the way that Dr. Chu-Tan and Dr. Mustapha Kamil present their research in their 3-minute talks.

The two problems introduced in Act 4 relate to the difficulty of obtaining permission to sample the drill core. These are relatively minor issues from a scientific point of view but were major decisions for English Heritage. Their import can be emphasized to add interest.

The above summaries provide only the barest outlines of the presentations. We did not summarize any of the data. When the talks are presented much more time would be spent on the data than on the questions or conflicting evidence. However, without these punctuations the presentations would be very dry. They would be unlikely to hold the audience's interest. The addition of questions and conflicting evidence (BUTs) gives you a way of structuring your talk. Adding them requires you to think about how you are going to present your data so that the presentation does not become a long list of graphs and charts. Your audience will benefit by their addition.

Your Turn—Practice with the Five-act Structure

Review the summary of Drs. Orth and McGlathery's work on eelgrass given below then put it into a five-act structure. The first act can be the same as in the three-act structure from Chapter 2. The other acts should be revised. Be sure to add at least one BUT in Acts 3 and 4. There are only a few problems described in this project, so it may not be possible to find a major conflict to begin Act 4. Try your own formulation before you read ours, which is given at the end of the chapter.

Restoration of seagrass habitat in coastal Virginia

In the 1920s, the marine slime mold *Labyrinthula zosterae* (eelgrass blight) began wiping out eelgrass (*Zostera marina*) beds in the coastal waters and lagoons of Virginia (USA). By the mid-1930s the plant was virtually extinct in the Virginia lagoons. With its disappearance water turbidity increased, which led to the hypothesis that decreased light quality was responsible for the lack of eelgrass recovery. However, in the late 1990s a natural patch of eelgrass was discovered, probably from seeds hitchhiking on recreational boats. Subsequent studies showed that light was not the limiting factor preventing spread of the plants.

Because of this, Drs. Orth and McGlathery hypothesized that the lack of recovery was due to the lack of a seed source rather than light quality. To test this hypothesis, they established test plots of eelgrass in the lagoons. Between 2001 and 2004, approximately 24 million seeds were broadcast into experimental plots in four coastal lagoons having no seagrass. Checks of the test beds in spring showed a 5–10% germination and establishment rate for the new plants. Aerial photographs showed 38% of the area of the test plots was covered by eelgrass. These plots demonstrated that seed supply rather than water quality was the limiting factor in seagrass recovery (Orth *et al.*, 2006). To determine the best areas to expand the project, the team combined field measurements of light availability with a modeling approach to identify 200 ha suitable for eelgrass restoration (Lawson *et al.*, 2007). The team then enlisted volunteers to broadcast seeds into these areas. Seeds for the large-scale project were first collected by hand, but the scale of the project soon made this impossible. An eelgrass harvesting machine was developed to cut the plants and remove the seeds. Once removed, the seeds had to be sorted.

– eelgrass –

The team discovered that viable seed sank faster than inviable seed, so they used a flume to separate the seeds. Good seeds fell closer to the water source and could be collected for sowing. Over the next 16 years volunteers spread more than 74 million seeds into 536 restoration plots totaling 213 ha. Turbidity initially declined with the establishment of the new beds, then increased, but the increased levels were still well below those observed before the reintroduction of eelgrass. Long-term monitoring of the restored beds showed that 3612 ha of vegetated bottom had been restored. A restoration program for bay scallops was begun, which resulted in a sustained population of scallops. The program has restored a remarkably hardy ecosystem that is trapping carbon and nitrogen; carbon that would otherwise contribute to global warming and nitrogen that would otherwise diminish water quality. It is one of the most successful marine restoration efforts in the world (McGlathery *et al.*, 2012; Orth *et al.*, 2020).

Structures for Longer Presentations

The presentations we have worked with up to this point should require no more than 20 minutes to present. For longer presentations we can add more acts to contain additional results, but we need to do this in a way that holds the audience's interest. The goal is to create a series of self-contained stories that, when taken together, explain your research (Craig and Yewman, 2014).

How you do this will depend heavily on your research area, the expectations of your audience, and your data. Each story will begin with a problem statement (a BUT) and will contain the data that addresses this problem. The issues that are still unresolved will form the problem statement of the next act. Each act will focus on a single major question and its answer.

Neurotoxic peptides from the giant Australian stinging tree

Let us return to Dr. Vetter's study of *Dendrocnide* stings as an example. We will keep Acts 1, 2 and the last act the same as in our previous five-act summary. The intervening acts will focus on the major questions of the research. Our summary of these acts only includes the problem statements that will form the backbone of the presentation. You would elaborate each of these acts if you were giving the talk.

Act 1: Contact with *Dendrocnide* plants causes excruciating long-lasting pain. The causative agent may be moroidin.

Act 2: *D. excelsa* trichomes lack moroidin; purified extracts of moroidin do not produce the same long-lasting effects as the stings; injection of trichome extract purified to exclude moroidin still produced the symptoms of a sting, so other pharmacologically active compounds must be responsible for the stings.

Act 3: BUT = Can we isolate the pharmacologically active compounds?

Act 4: BUT = Are the peptides we isolated responsible for the stings?

Act 5: BUT = How do they cause the stings? Do they act preferentially on neurons?

Act 6: BUT = Are they neurotoxins? Do they act directly on Na_V channels?

Act 7: BUT = Do they localize to the trichomes?

Act 8: Final conclusions—The causative agents are neurotoxins similar to those found in spider and cone snail venom.

The trick to making a longer presentation interesting is to present the questions as problems that need solution. To do this you introduce each act with a brief summary of what has already been presented and then introduce a problem that still needs to be solved. For instance, Act 4 could be introduced in this way, "We now had good candidates for the causative agents, but we still needed to show that these peptides could induce long-lasting stings." Act 5 could be introduced like this, "So we had good evidence that the peptides were causing the stings. But how do they do this?" Breaking the talk into a series of smaller stories, each beginning with a problem and ending with its solution will keep the audience interested to the end of the talk.

The origins of the Stonehenge megaliths

Let us look at one more example. Professor Nash's work on Stonehenge can easily be expanded to an hour-long talk. Our outline follows.

Act 1: The sandstone megaliths of Stonehenge are thought to have originated from the Marlborough Downs, but there has been no scientific test of this hypothesis.

Act 2: A meter-long drill core from Stone 58 is returned to English Heritage after being missing since 1958.

Act 3: BUT = Can the provenance of the silcrete rocks be established?

Act 4: BUT = Do the 52 remaining megaliths all have the same chemistry? Do they all come from the same area?

Act 5: BUT = Can the new provenancing method be used to find the location of Stone 58 and, by implication, the other 49 megaliths with the same chemistry?

Act 6: Final conclusions—Stone 58, and by correlation 49 other megaliths, is localized to an area on the south-eastern edge of the Marlborough Downs.

The results of the geological thin sections and chemical analyses can be presented in Act 5 as a supplement to the mass spectrometry data. Results like these can be presented as small scenes within the larger act. Like the act itself, they should begin with a problem and end with its solution.

Each act addresses a problem. Each has its own beginning, middle and end. What the audience learns in one act provides the information needed to understand the next. The same is true for the scenes. They each begin with a problem, proceed through the relevant evidence, and end with a conclusion. Each conclusion is tentative until you reach the end of the talk. The final act ties everything together and answers the question posed at the beginning. The symmetry between the beginning and end gives a sense of completion.

Slides to Enhance Longer Presentations

Just as your talk is divided into a series of acts, your data should be divided into simple, natural units and placed on separate slides. To achieve this, it is best to prepare your remarks first, then your slides (Craig and Yewman, 2014). Find the flow of your talk then add visuals. The narrative flow should be like the flow of a river. It should move steadily in one direction. Each slide should move you slightly farther downstream. When there are multiple pieces of information on a slide, they easily create eddies in the movement. You get stuck in a small whirlpool and momentarily lose sight of where you are going. Ideas flow well when each slide presents a single idea and does not contain too much text. If you force yourself to change slides for each point it will be easier to notice when the narrative flow is disrupted. You can then make a conscious decision about whether that disruption is necessary or whether you should eliminate it in preference to continuing the main flow.

As we learned in Chapter 3, text and images on the same slide can be difficult to interpret. Simple photographic images with no text are clearer and easier to understand. However, few technical presentations can rely solely on photographs. Some text is necessary, as are graphs and other figures. But remember, listening is not like reading. A listener cannot stop the talk if they get confused. They cannot speed it up or slow it down. They will become frustrated when there is too much text. They will not be able to listen and read at the same time. To adjust they will stop listening (Lebrun, 2009). This is not what you want. You want to be the center of attention. You want to present the data. You do not want to be the background noise while your audience reads.

Like too much text, placing multiple graphs or images on a slide will confuse your audience. Remember that they are seeing your data for the

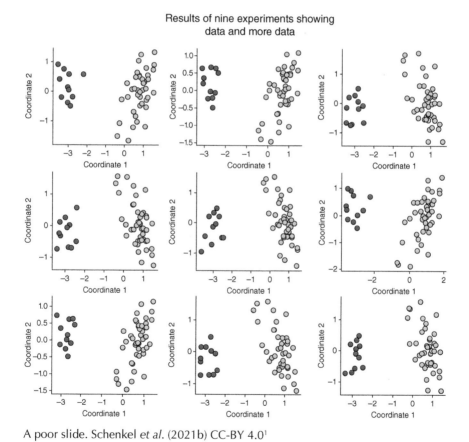

A poor slide. Schenkel *et al.* (2021b) CC-BY 4.0[1]

---

first time. They will not easily understand a complex slide. Look at the two examples. One is poor. The other is better. There are too many graphs on the poor slide, and the title is too long. The complexity of the data means that the audience will spend most of their time trying to interpret the graphs. This will prevent them from understanding the data and will interfere with your verbal points. The second slide is much better. There is a single graph, and the title is shorter. Because there is less data the title can provide a better summary. With this slide the audience can both listen to your presentation and assimilate the data. There is still the problem of having text and images on the same slide, but this design minimizes the dissonance. Keep your slides simple and focused on one main point. Your audience will thank you for it. Carter (2013) has many other excellent recommendations for slide design.

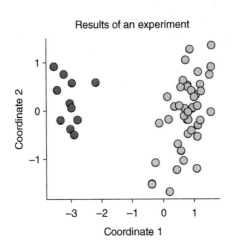

A better slide. Schenkel *et al.* (2021b) CC-BY 4.0[2]

In closing, we want to make it clear that the authors of the figures in this section (Schenkel *et al.*, 2021a) never intended them to be presented in this way. We have modified their figures to illustrate our points. We have also changed the colors of the data points to adapt them for people living with color blindness. There are excellent resources on the web to help you choose appropriate colors (Flück, 2006–present).

Answer to Exercise

Restoration of seagrass habitat in coastal Virginia

Five-act structure

Act 1: **current state of knowledge**
  a. Eelgrass blight wiped out eelgrass in the Virginia lagoons. With its disappearance water turbidity increased. Increased turbidity led to the hypothesis that decreased water quality was responsible for the lack of eelgrass recovery.

Act 2: **conflicting evidence; new hypothesis**
  a. **BUT:** Natural recovery of eelgrass in one of the lagoons and studies that showed light was not the limiting factor called the turbidity hypothesis into question; test plots established spreading colonies with a 5–10% germination and establishment rate (**conflicting evidence**).
  b. Lack of seed rather than water quality is responsible for the lack of eelgrass recovery (**new hypothesis**).

Act 3: **first presentation of results**
  a. The team uses field measurements and modeling to identify 200 ha suitable for eelgrass restoration; a large-scale reseeding project is established.
  b. **BUT:** The project is so large that it becomes impossible to harvest seeds by hand.
  c. An eelgrass harvesting machine is developed and a flume is used to separate viable from inviable seeds.

Act 4: **problems and their solution**
  a. Over the next 16 years a team of volunteers spreads more than 74 million seeds over 213 ha of the lagoons.
  b. **BUT:** Turbidity increases after initially declining.
  c. Fortunately, turbidity remains below its level before the reintroduction of eelgrass.
  d. Other restoration programs are begun, including one for bay scallops.

Act 5: **conclusions**
  a. The project has restored 3612 ha of the lagoons to eelgrass beds. Bay scallops have been found in aquaculture beds up to 20 km away from where they were set out.

references

Carter, M. (2013) *Designing Science Presentations: a Visual Guide to Figures, Papers, Slides, Posters, and More*. Academic Press, San Diego, California.

Craig, A. and Yewman, D. (2014) *Weekend Language: Presenting with More Stories and Less PowerPoint*. DASH Consulting, Portland, Oregon.

Flück, D. (2006–present) Colblindor. Available at: https://www.color-blindness.com/

Gilding, E.K., Jami, S., Deuis, J.R., Israel, M.R., Harvey, P.J., Poth, A.G., Rehm, F.B.H., Stow, J.L., Robinson, S.D., Yap, K., Brown, D.L., Hamilton, B.R., Andersson, D., Craik, D.J., Vetter, I. and Durek, T. (2020) Neurotoxic peptides from the venom of the giant Australian stinging tree. *Science Advances* 6: eabb8828. doi: https://doi.org/10.1126/sciadv.abb8828

Lawson, S.E., Wiberg, P.L., McGlathery, K.J. and Fugate, D.C. (2007) Wind-driven sediment suspension controls light availability in a shallow coastal lagoon. *Estuaries and Coasts* 30, 102–112. doi: https://doi.org/10.1007/BF02782971

Lebrun, J.-L. (2009) *When the Scientist Presents: an Audio and Video Guide to Science Talks*. World Scientific Publishing, Singapore.

McGlathery, K.J., Reynolds, L.K., Cole, L.W., Orth, R.J., Marion, S.R. and Schwarzschild, A. (2012) Recovery trajectories during state change from bare sediment to eelgrass dominance. *Marine Ecology Progress Series* 448, 209–222.

Nash, D.J., Coulson, S., Staurset, S., Ullyott, J.S., Babutsi, M., Hopkinson, L., and Smith, M.P. (2013) Provenancing of silcrete raw materials indicates long-distance transport to Tsodilo Hills, Botswana, during the Middle Stone Age. *Journal of Human Evolution* 64, 280–288. doi: https://doi.org/10.1016/j.jhevol.2013.01.010

Orth, R.J., Luckenbach, M.L., Marion, S.R., Moore, K.A. and Wilcox, D.J. (2006) Seagrass recovery in the Delmarva Coastal Bays, USA. *Aquatic Botany* 84, 26–36. doi: https://doi.org/10.1016/j.aquabot.2005.07.007

Orth, R.J., Lefcheck, J.S., McGlathery, K., Aoki, L., Luckenbach, M.W., Moore, K.A., Oreska, M.P.J., Snyder, R., Wilcox, D.J., and Lusk, B. (2020) Restoration of seagrass habitat leads to rapid recovery of coastal ecosystem services. *Science Advances* 6: eabc6434. doi: https://doi.org/10.1126/sciadv.abc6434

Schenkel, L.C., Aref-Eshghi, E., Rooney, K., Kerkhof, J., Levy, M.A., McConkey, H., Rogers, R.C., Phelan, K., Sarasua, S.M., Jain, L., Pauly, R., Boccuto, L., DuPont, B., Cappuccio, G., Brunetti-Pierri, N., Schwartz, C.E., and Sadikovic, B. (2021a) DNA methylation epi-signature is associated with two molecularly and phenotypically distinct clinical subtypes of Phelan-McDermid syndrome. *Clinical Epigenetics* 13: 2. doi: https://doi.org/10.1186/s13148-020-00990-7

Schenkel, L.C., Aref-Eshghi, E., Rooney, K., Kerkhof, J., Levy, M.A., McConkey, H., Rogers, R.C., Phelan, K., Sarasua, S.M., Jain, L., Pauly, R., Boccuto, L., DuPont, B., Cappuccio, G., Brunetti-Pierri, N., Schwartz, C.E., and Sadikovic, B. (2021b) Additional file 3 of DNA methylation epi-signature is associated with two molecularly and phenotypically distinct clinical subtypes of Phelan-McDermid syndrome. Available at: https://doi.org/10.6084/m9.figshare.13533178.v1

Poster Presentations 6

Posters are the most difficult of all scientific presentations. A poster has to fulfill three essential yet somewhat contradictory purposes. It must attract passers-by so that potential visitors can determine if they are interested in your work. It must serve as the backdrop for your oral presentation. Finally, if it will be on display outside of your presentation time, it must stand on its own and present the full content of your research. At least two of these purposes are in direct conflict. A good deal of text is necessary for a poster to stand on its own and represent your research. However, too much text impedes a visitor's ability to quickly understand your results, and too much text can get in the way of your oral presentation. The art of preparing a good poster requires you to find a way to resolve these conflicts (Faulkes, 2021).

Writing a Good Title

The best way to help someone understand your poster is with your title. Your title should concisely state your main result. It should communicate your result clearly, and with minimal jargon. If you are writing for a scientific audience, you can use technical terms but remember that scientists outside of your specialty will not be familiar with your discipline's jargon. For instance, most ecologists will easily understand the term "acid sulfate soil" even if they do not know the details of how this term is used

in Australia, but they are likely to be stymied by the abbreviation ASS. The words "acid sulfate soil" are technical but can be understood by most ecologists. ASS is jargon. It will most easily be understood by someone working on these soils. For an audience in your research area you can use technical terms but even here it is best if you avoid jargon. Of course, if you are presenting to the general public even technical terms should be avoided.

Titles that are simple, surprising, concrete, and that avoid jargon are always best no matter what audience you are addressing. Let us look at some examples. The titles we have selected for analysis all come from first-class journals and meet our criteria of being simple, surprising, and concrete. They clearly state the results of the research and draw the reader in. They make them want to know more. Although we suggest changes to illustrate how even these good titles can be improved, we do not mean to impugn their quality. We have selected them because they are so good. Although the titles come from journal articles, they could easily be used for posters, which is how we intend them here.

Teosinte ligule allele narrows plant architecture and enhances high-density maize yields (Tian *et al.*, 2019)

This is a good title but let us see if we can improve it for use with an audience outside your discipline. The words "high-density" are a bit unclear. Does "high-density" refer to maize, to yields, or to both? What is a high-density maize yield? We should be able to answer these questions without reading the poster. Let us look at what happens if we remove the words high-density and add the indefinite article to the beginning of the title: *A teosinte ligule allele narrows plant architecture and enhances maize yields.* This is better. The use of the indefinite article makes it clear that the word "teosinte" qualifies the origin of the allele. It comes from the wild ancestor of domestic maize, teosinte. The removal of "high-density" also clarifies the meaning. The poster will further clarify this by explaining that the improvement in yield is only seen when maize is grown at high planting densities. This does not have to be explained in the title. The role of the title is to draw the reader in, not to explain every aspect of the research. The title is now more striking, but it can still be improved. Let us see what happens when we clarify the meaning of "narrower plant architecture." *Introducing a teosinte allele into maize produces narrower plants and increases yields.* In the original title "architecture" is used as a technical term. Only maize specialists and plant morphologists will fully understand it. Removing it allows the title to appeal to a wider audience. Moving "maize" closer to the beginning of the title also quickly tells the reader what crop is under study. The title is now simple, clear, and at least somewhat surprising. The reader is more likely to stop and explore the poster. Once they do, you can present the limitations of your research. In this case, the beneficial effect of the allele is only seen when maize is planted at high densities. Under these conditions the narrower plants produce more corn. There is no effect at low planting densities. This makes perfect sense now that we understand that the genetically modified plants are narrower. If the plants are spaced far apart, there is no benefit of the narrower architecture. These subtleties are important, but do not need to be in the title.

Increasing wildfires threaten historic carbon sink of boreal forest soils (Walker *et al.*, 2019)

This is a great title that alerts the reader to the importance of the research and hints at its major conclusions. Because of its importance for climate change, the authors focus on the importance of the work rather than their specific results. Their results are based on work done after the 2014 wildfires in Canada's Northwest Territories and are narrowly applicable only to those forests, and only to forests of a certain age (< 60 years old). These are important facts, but they do not need to be part of the title, especially when the work is of such topical importance. The authors want to draw attention to the fact that the burning of legacy carbon (carbon that has accumulated in the soils from previous fires) could release unexpected amounts of CO_2. The authors have rightly chosen to focus on this important conclusion rather than their narrow findings. Of course, their results must be replicated in other forests, but this is always the case and does not need to be emphasized in the title. The limitations of the research and the justification for extending their conclusions beyond their study can be presented in the body of the poster. If we wanted to increase the legibility of the title for a more general audience, we might consider *Increasing wildfires threaten the boreal forest's role as carbon sinks*. This title is slightly less nuanced, but still accurate, and is a bit easier to understand.

Two new species of shrew-rats (*Rhynchomys*: Muridae: Rodentia) from Luzon Island, Philippines (Rickart *et al.*, 2019)

Taxonomic posters present a unique set of problems. The core message is the taxonomic descriptions, which are much too long to summarize in a title. The example we have chosen does a nice job of presenting the names of the new species at several different levels of the taxonomic hierarchy (*Rhynchomys*—genus; Muridae—family; Rodentia—order), and the geographical occurrence of the new species (Luzon Island, Philippines). The title would be even better if it included the specific epithets of the new species. This would add only

two words and could look like this *Two new species of shrew-rats (Rhynchomys labo, R. mingan: Muridae: Rodentia) from Luzon Island, Philippines*. This type of explicitness is particularly useful in poorly known groups where there are likely to be multiple papers each describing only one or two new species. In cases of major revisions, explicitly naming the species will be impossible. However, major revisions are seldom presented as posters. Whenever possible, you should write explicit titles.

Habitat divergence shapes the morphological diversity of larval insects: insights from scorpionflies (Jiang *et al.*, 2019)

This is an excellent technical title and is about as good as we can do given the nature of this research and the constraints of writing a short, simple title. The paper concludes that while larval anatomical characteristics define monophyletic groups, body coloration and external characteristics do not. Scorpionfly larvae that live in the soil have pale trunks, reduced antennae, shortened setae, and flattened compound eyes, all adaptations to life underground. Larvae that live above ground are dark on their upper surfaces and possess well-developed appendages that can be used for movement. Both characters are adaptations to avoid predation. The title does a good job of presenting these conclusions while making it clear which organisms were studied. The only problem is that the title is not very concrete. What type of insights were gained from scorpionflies? Are they specific to scorpionflies, or are scorpionflies an example of a general pattern? We can try to clarify this, but the complexity of the results makes it difficult: *The larval anatomical characteristics of scorpionflies are correlated with phylogeny, but their external morphology is better explained as an adaptation to the environment.* This captures the central message but is wordy, and it narrows the results to scorpionflies. Sometimes there is a trade-off between simplicity and concreteness. The paper's title is simple but not concrete. Our rewording is concrete, but not simple. If the authors believe that their conclusions have validity beyond scorpionflies they could solve these problems by shortening the title to: *Habitat divergence shapes the morphological diversity of larval insects.* Then the body of the poster would introduce scorpionflies and explain the details of the research and justify their conclusion that their results are valid for other larval insects. With slight modification this shortened title could also be used for a general audience. If we remove the technical terms we have: *Changes in habitat influence the external appearance of larval insects.* Which of these titles works best will depend upon your audience.

Some closing thoughts on titles

Although your title should state your major conclusions, it should not overstate them. Your title must be easy to understand but you should not imply wide applicability if your results restrict their validity. There is a fine line here. You must be completely honest in all aspects of your presentation, but honesty does not mean that every detail of your research must appear in the title. Unless you believe the limitations of your research place important restrictions on the validity of your findings, you should deal with them in the body of the poster.

If your results depend heavily on modeling, you should make this clear in your title. Modeling studies are extremely valuable, but they are different from empirical studies. This difference should be made clear.

Your turn—writing titles

Read the following abstracts and see if you can improve the titles. Our formulations are at the end of the chapter.

Hermaphroditism promotes mate diversity in flowering plants

ABSTRACT

PREMISE: Genetically diverse sibships are thought to increase parental fitness through a reduction in the intensity of sib competition, and through increased opportunities for seedling establishment in spatially or temporally heterogeneous environments. Nearly all research on mate diversity in flowering plants has focused on the number of fathers siring seeds within a fruit or on a maternal plant. Yet as hermaphrodites, plants can also accrue mate diversity by siring offspring on several pollen recipients in a population. Here we explore whether mate composition overlaps between the dual sex functions, and discuss the implications for plant reproductive success.

METHODS: We established an experimental population of 49 *Mimulus ringens* (monkeyflower) plants, each trimmed to a single flower. Following pollination by wild bees, we quantified mate composition for each flower through both paternal and maternal function. Parentage was successfully assigned to 240 progeny, 98% of the sampled seeds.

RESULTS: Comparison of mate composition between male and female function revealed high mate diversity, with almost no outcross mates shared between the two sexual functions of the same flower.

CONCLUSIONS: Dual sex roles contribute to a near doubling of mate diversity in our experimental population of *Mimulus ringens*. This finding may help explain the maintenance of hermaphroditism under conditions that would otherwise favor the evolution of separate sexes.

(Christopher *et al.*, 2019)

Inbreeding reduces long-term growth of Alpine ibex populations

ABSTRACT

Many studies document negative inbreeding effects on individuals, and conservation efforts to preserve rare species routinely employ strategies to reduce inbreeding. Despite this, there are few clear examples in nature of inbreeding decreasing the growth rates of populations, and the extent of population-level effects of inbreeding in the wild remains controversial. Here, we take advantage of a long-term dataset of 26 reintroduced Alpine ibex (*Capra ibex ibex*) populations spanning nearly 100 years to show that inbreeding substantially reduced per capita population growth rates, particularly for populations in harsher environments. Populations

with high average inbreeding (F≈0.2) had population growth rates reduced by 71% compared with populations with no inbreeding. Our results show that inbreeding can have long-term demographic consequences even when environmental variation is large and deleterious alleles may have been purged during bottlenecks. Thus, efforts to guard against inbreeding effects in populations of endangered species have not been misplaced.

(Bozzuto *et al.*, 2019)

Embolism resistance in stems of herbaceous Brassicaceae and Asteraceae is linked to differences in woodiness and precipitation

ABSTRACT

BACKGROUND AND AIMS: Plant survival under extreme drought events has been associated with xylem vulnerability to embolism (the disruption of water transport due to air bubbles in conduits). Despite the ecological and economic importance of herbaceous species, studies focusing on hydraulic failure in herbs remain scarce. Here, we assess the vulnerability to embolism and anatomical adaptations in stems of seven herbaceous Brassicaceae species occurring in different vegetation zones of the island of Tenerife, Canary Islands, and merged them with a similar hydraulic-anatomical data set for herbaceous Asteraceae from Tenerife.

METHODS: Measurements of vulnerability to xylem embolism using the *in situ* flow centrifuge technique along with light and transmission electron microscope observations were performed in stems of the herbaceous species. We also assessed the link between embolism resistance vs. mean annual precipitation and anatomical stem characters.

KEY RESULTS: The herbaceous species show a 2-fold variation in stem P50 from -2.1 MPa to -4.9 MPa. Within *Hirschfeldia incana* and *Sisymbrium orientale*, there is also a significant stem P50 difference between populations growing in contrasting environments. Variation in stem P50 is mainly explained by mean annual precipitation as well as by the variation in the degree of woodiness (calculated as the proportion of lignified area per total stem area) and to a lesser extent by the thickness of intervessel pit membranes. Moreover, mean annual precipitation explains the total variance in embolism resistance and stem anatomical traits.

CONCLUSIONS: The degree of woodiness and thickness of intervessel pit membranes are good predictors of embolism resistance in the herbaceous Brassicaceae and Asteraceae species studied. Differences in mean annual precipitation across the sampling sites affect embolism resistance and stem anatomical characters, both being important characters determining survival and distribution of the herbaceous eudicots.

(Dória *et al.*, 2018)

The Content of Your Poster

We now turn our attention to the content of your poster. The size of scientific conferences and the limitations of the poster format make it impossible to present all aspects of your work. You are going to have to focus on something. What will it be?

If your title is concise and if it summarizes your main result, you will be drawing people in to hear more. What do you want them to remember? If you are smart, you want them to remember what is original about your work, and you want them to remember you. Most of all you want them to remember you. If they remember you then, when they have a job opening they may think of you and invite you to apply. How great would that be? As a young scientist, remembering you is the most important thing a visitor can take away from your poster.

To get them to remember you, you will need to tell them about what you have accomplished, what is unique about your work. You should design your poster with this in mind. No matter whether the primary use for your poster will be as a backdrop to your verbal presentation or whether you want someone to be able to understand your work by reading it when you are not present, you need to design your poster to focus on your results. One good way to do this is to use the central portion of the poster, where your visitors' eyes are naturally drawn, to present a concise summary of your work. In the best designed posters this summary will be presented graphically through images, graphs, and charts. A visitor should be able to look at this area and gain a basic idea of what your work is about. Then, if you have piqued their interest, they can look to the rest of the poster to gain a deeper understanding. To facilitate this understanding, the text should be presented in easily understood bullet points. Paragraphs require the reader to extract the important information from the text. This requires mental effort and increases the viewers' cognitive load. It makes it harder for them to concentrate on your main point. They are less likely to remember you if they must spend a good deal of mental energy deciphering your text. If they can look at your poster and quickly understand your main point then they will have more energy to remember your research. Good bullet points do most of your viewers' thinking for them. They summarize the important aspects of your research and present it in an easily digestible form.

Unless your poster is about the development of a new method, the majority of your methods should be placed elsewhere. They can be placed on a website linked to by a quick response (QR) code or handed out on a separate sheet. QR codes are two-dimensional, matrix barcodes that are machine readable with a smartphone. They are commonly used to encode website URLs. Scanning the code opens the linked website. By including a QR code you can link to a website that contains your methods and additional information about your research. If you have established tracking metrics on your site the inclusion of a QR code will allow you to determine how many people follow up on your poster. QR codes can be created for free on many websites designed for this purpose. An internet search will yield several alternatives.

If you feel your methods are essential to your presentation and must be on your poster, you can place them in a separate section at the bottom of the poster, in a smaller font size. This is how the best journals handle methods. In *Nature* and *Science* the methods are abbreviated, placed at the end of the text in a separate section, and set in a smaller font size. This sends the message that the methods are there for scientists working on closely related problems. No one else needs to know them in detail. Your results are much more important. Review of the manuscript has assured that the methods are accurate and appropriate. You should treat the methods on your poster in the same way. That way you can focus on your results and only discuss your methods if someone has a question. You are not looking for a critique of your methods. Try not to get yourself into that situation. Make sure your methods are correct, then ignore them during your presentation. If you are a graduate student, your advisor/supervisor and committee will already have reviewed and critiqued your methods. If you are a postdoc or a faculty member, you should be confident in your methods before you present. Focus on your findings. Your audience will appreciate it.

Just like your methods, it is less important that your poster explains the background of your research than it explains your results. The background is based on work that was done by other scientists. If your audience is in your field, they probably already know this work. If they are not, you can summarize it quickly without going into great detail. If you are a student, your advisor/supervisor has probably insisted that you do a lot of work to understand the background to your research. They asked you to do this to assure that you understand how your research builds on what has already been done. You are now at the stage where you are ready to present your own work. You have earned your right to be here. You do not have to prove you know the background. Focus on what you have accomplished not on what others have done. Only include the minimal amount that your visitors need to know to understand your work.

Look at the abstract of the paper *Hermaphroditism promotes mate diversity in flowering plants* (Christopher *et al.*, 2019) that we reproduced earlier in the chapter. The work reports an important finding, but it is only important in the context of previous work. This paper is a great example of how to handle background information. The authors deal with the previous research in one sentence by saying "Nearly all research on mate diversity in flowering plants has focused on the number of fathers siring seeds within a fruit or on a maternal plant." It is a short summary but is extremely important. Until the authors explained this context, their work seemed self-evident. Of course, organisms that possess both sexes will have more mating partners than those which are unisexual. How could it be otherwise? This result only becomes interesting when we know about the previous work that has been done on mating systems. When we know how that work was done, work that tracks both types of parental contributions seems novel and exciting. The background work is important, but we do not need to know the details. This is how you should treat the background to your research.

Based on everything we have said so far you can probably figure out what we recommend about acknowledgments and references. In terms of your primary goals neither the acknowledgments nor the references are important. Depending on your discipline, they may both need to be present, but neither should be emphasized. You certainly should not speak about them when you are presenting your research. They are there only in case someone wants to glance at them when they are reading your poster. If you include them, put them at the bottom in a smaller font. Even better, include a QR code with a link to these sections on a website.

We hope by now that you see that your poster should not have an abstract. Your poster itself is your abstract. It presents the main findings of your research in a concise and easily digested form. Why would you want an abstract on an abstract? The only reason to include one would be if you designed your poster so poorly that the average reader could not understand it. In that case they might need to read your abstract for guidance. Unfortunately, many scientific posters are like this. They are so hard to understand that an abstract is needed to explain the research. Design your posters so they are easy to understand. That way you will not need an abstract and you will have more space to present your results.

This brings us to the main part of your poster, your results. The best way to present your results is with images, graphs, and charts. You should select the most appropriate graphics, arrange them in the center of your poster, and provide just enough text to clear up any ambiguities. The text should be presented in bullet points, not paragraphs. Potential visitors will be attracted to your graphics. From there they will read your title, and if it interests them, they will stop and read the text. They are unlikely to stop if they see a lot of text. This is especially true if they find a title that does not convey your major finding. Your text should not be the primary focus of attention. Visitors should be able to skip most of the text and still understand your points. The text only becomes important after a visitor has decided to stop and learn more. The graphics, supported by a good title, will draw them in. The text will give them a deeper understanding. If you are lucky, you may encounter someone who has deep interest in your work and wants to know more. A QR link to your published paper, or to more information about your research, will maintain engagement after the poster session ends.

Now let us suppose you are standing in front of your poster during the formal poster session and someone approaches to learn about your work. In this situation the graphics and title play the same role they do outside of the poster session, but now the text becomes irrelevant. Because you are present to explain your work the text becomes a distraction that takes up space on the poster that could better be used to display graphs and charts. The more you can use images, graphs, and charts to support your verbal presentation the better your visitors will understand your work.

Let us say you are in the middle of explaining your work and someone new walks up to your poster. What should you do? Should you ignore them and continue speaking to those who arrived first? Should you start over from the beginning? The most common behavior we have seen is for the speaker to ignore the newcomer and continue their presentation. But what if that new person was approaching you because they had a job opening and wanted to learn more about your work? In  that case, ignoring them could be a real mistake. If they are exceptionally persistent they might wait until you are done with your presentation to speak to you, but why take the chance? Why not welcome them and include them in the conversation? When the new person approaches turn slightly toward them so that your body language shows that you are clearly speaking to two people. If it seems appropriate, nod at them to indicate that you have recognized their presence. Then welcome them verbally and very briefly summarize what you have been talking about. You want to give them a brief introduction to the conversation so that they can follow it when you pick up where you left off. They will not receive all the details, but they should be able to follow your main line of argument. You want to make sure that they feel included in the conversation. Once the first person has left, you can turn your full attention to your new visitor and fill in any details that they missed. This procedure will assure that you do not lose the interest of either the earlier, or the later visitor. You can repeat this process several times as new visitors arrive, although admittedly it becomes difficult after the third arrival or so. At some point you will have to welcome the new visitors with changes in body language and gestures, but without stopping your presentation. Keep these presentation requirements in mind as we go on to talk about poster design.

Poster Layout and Design

Posters are not the place to present everything about your research. Posters are visual summaries of your most important results. Looking at a poster from a distance is like looking at an image. It is not like reading a book. When looking at a poster our eyes are first drawn to the most striking elements (Dahal, 2011). These are the images, graphs, and charts. If these are presented in color, they will more readily draw the viewer's eye. We will see this clearly when we look at the results of eye-tracking studies in the next chapter.

The most important part of your poster is the center. A visitor's eyes will first be drawn to the center, especially if it contains a graphic element. You can force your viewers to look at other parts of the poster by placing design elements in these positions, but their eyes will naturally be drawn to the center. You can work with this natural tendency by placing your most important figures and graphs in the center.

Before we look at examples of good designs let us consider some elements that should be on all posters. The title and the authors' names and their affiliations all need to be given a prominent place. The title should be set in a large font size and placed in a prominent position. Most poster designs place it across the top, but as we will see other positions are possible. The authors' names should be set in a smaller font size. Their most common position is immediately below the title, but again other positions are possible. While the authors' affiliations need to be included, they are much less important than the title and the authors' names, and so should be set in a still smaller font size. It is most common to place these affiliations immediately below the authors' names but there is no reason they must be in this position. The QR code, if you include one, often creates a small dead space next to it that could be used for this purpose.

QR codes are becoming increasingly common on posters and you should consider including one. As the codes are relatively innocuous, they can be placed almost anywhere on the poster. We have suggested a location for one on each of the designs we discuss but other positions are possible. We have even seen posters where the central graphic was replaced by a large QR code, though we would not say that this is the best use of the center.

Many scientists like to place their university's logo on their posters, often in a prominent position. This is a nice feature, but we do not feel that is it essential and in some situations it can become problematic. If there are multiple authors all from different institutions, dealing with the multiple different logos can take up a good deal of space. Even when the authors are from a single institution, giving a strikingly bold logo a prominent position on the poster can cause the viewers' attention to be directed at the logo instead of the body of the poster. We will see an example of this when we look at the eye-tracking simulations in the next chapter. If you are going to include the logo, place it in a less prominent position, especially if it is a bold, strikingly colored logo.

Rather than including your school's logo, you should consider including a high-quality headshot of yourself as the first author. You should use a photograph that shows you in the best possible light; a photograph from which you can be easily recognized, not a photograph that shows off some quirky aspect of your personality. You want other scientists to be able to easily recognize you from your photograph. Then, if they only know you from your picture, they will be able to recognize you at a coffee break. If they recognize you, they are much more likely to stop and introduce themselves, which is exactly what you want. You have succeeded in getting them to remember you. Congratulations.

Examples of good designs

We will begin our consideration of poster layouts with a discussion of Mike Morrison's work to create a Better Poster (Morrison, 2021). Morrison recommends focusing on your key results, including explanatory graphics and reducing the amount of text (see Morrison, 2021). He also suggests the inclusion of a QR code linking to a website with additional resources. The center of the poster is reserved for a description of your most important result, written in plain English with a minimal amount of jargon.

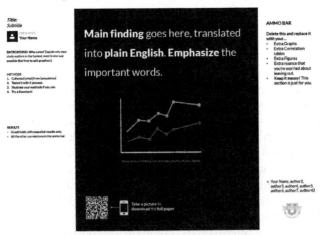

Template for a Better Poster (Morrison, 2021), CCO[1]

---

[1] https://creativecommons.org/publicdomain/zero/1.0/

This description can be supported by a key figure that supplements the result. The QR code is placed below these elements, at the bottom of the central section of the poster. A solid color is used for the background of this section.

The right and left sides of the poster present supporting information. Morrison describes the right side as the "Ammo Bar." It is a place to put the graphics you need to explain your research to visitors. The figures do not need to be self-explanatory as you will be present to explain them. A full list of the authors, their affiliations, and the schools' logos go at the bottom of this column. The left side of the poster contains a more traditional explanation of your results, including text and figures. This is the "Silent Presenter" bar. The information included here is intended to be read when you are not present. The left side also contains the name of the first author and the title, which is distinct from the description of your most important result in the center of the poster. However, you can use this description as your title, as we suggest earlier in the chapter. The Better Poster QR code links to Morrison's description of his idea.

Mike Morrison, Better Poster 1

Let us see how the Better Poster design fulfills the three requirements: (i) it must attract potential visitors; (ii) it must serve as a backdrop for your oral presentation; and (iii) it must stand on its own and present the full content of your research. The colored background with a clear presentation of your main finding in the center of the poster is a powerful attractant for visitors. The central graphic will reinforce this message and provide additional interest. The materials in the left column allow visitors to understand your work at a deeper level. If they want to know more, the QR code provides a link to additional information. As a backdrop to your presentation, the material in the right column provides what you need to talk about your work. Remember that your visitors will approach your poster already knowing your central finding because you have presented it at the center of the poster.

Of the three essential functions of a poster, the Better Poster design does a great job attracting attention and providing a background for your verbal presentation. It does a bit less well in telling your story all by itself, mainly because it uses so much of the central space to deliver your key message. The difficulty presenting your research when you are not present is a problem with many posters, and it is one to which you need to give considerable thought. How little can you include on your poster and still convey the most important details of your research? If you add too much text you will drive away visitors and make it more difficult to explain your research verbally. If you include too little, your visitors will find your work difficult to understand. If you are like most scientists, your tendency will be to add too much text. Be aware of this tendency and act accordingly.

We wish we could give you the magic formula that would fulfill the three conflicting requirements of a good poster, but there are discipline-specific issues that only you can address. We can draw attention to the problems, but you will have to find the solution that works for you. Explain as much as you can graphically and add only enough text so that the reader can understand what the graphics show. That is our best advice.

Since first proposing the Better Poster, Morrison has developed his ideas further and highlighted other good designs. The Better Poster 2 QR code links to a video where he presents these ideas.

Mike Morrison, Better Poster 2

Our next example of good design is a classic three-column design with plenty of space for a large graphic element in the center (left poster, below). The center of the poster should be devoted to a presentation of your key result, with additional details and supporting information in the side panels. There is plenty of space for a large title. If you need to present more than one graphic, there is space to include multiple images (right poster, below). With a design like this, and with a title that clearly states your key finding, you will attract attention. To prepare for visitors who want to discuss your results in more detail you can add a QR code that links to additional information. There is also space for a picture of the first author.

author photo	**Title** authors & affiliations		author photo	**Title** authors & affiliations	
Text & Graphics	**Graphics (supplemented by text)**	**Text & Graphics** QR	**Text & Graphics**	**Graphics (supplemented by text)** / **Graphics (supplemented by text)**	**Text & Graphics** QR

Based on Tressel and Vincent (2019), used with permission.

Our third example is a variation on the three-column design with a different division of the space and a different placement of the authors'

Authors	**Title**	
Text & Graphics	**Graphics (supplemented by text)**	Supplemented text & Graphics
Author affiliations		Author photo QR

Based on Stevens *et al.* (2019), used with permission.

names and affiliations. In this design the title is centered over the right two columns and the authors' names and affiliations are split and placed at the top and bottom of the left column, respectively. The left column is used to provide text and graphics, while the center of the poster presents your main results in graphs, charts, and images. The right column is used for additional information that cannot be included in the center. A picture of the author and a QR code can be placed at the bottom of the right column.

A three-column design can also be used in portrait orientation. The basic layout here is the same as the landscape designs. Your main result goes in the center with supporting information on the sides. A picture of the first author can be included in the top left corner and there is plenty of room for a QR code.

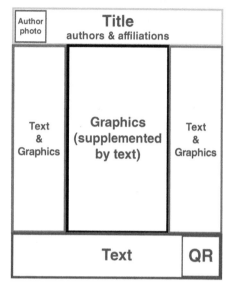

The last three designs we discuss divide the poster into a series of boxes, rather than columns. These designs can be used to present more detailed information. They are specialized designs that should only be used if your main result requires elaboration through the presentation of multiple pieces of evidence. If you can use one of the earlier designs, you should. If you decide to use one of these multi-box layouts, it is important to remember that your main result should be presented graphically and only supplemented with text.

Based on Jantzen *et al.* (2019), used with permission.

The purpose of the boxes is to present explanatory graphics, not text. The danger of these designs is that you will be tempted to fill the boxes with text, which will obscure your meaning. The amount of text should be kept to a minimum.

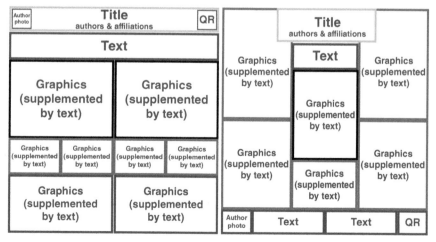

Based on Phillips *et al.* (2019) (left) and Ahlquist and Vincent (2019) (right), used with permission.

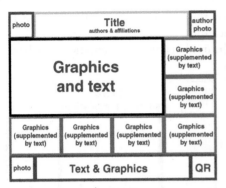

Based on Barker *et al.* (2012), CC-BY 4.0[2]

A second problem with multi-box designs is that they tempt you to include more results than can realistically be accommodated on a poster. Because there are many boxes it is tempting to include a different result in each box. This will quickly overwhelm your viewers and will decrease the impact of your work. Your poster must have a central message. During your presentation you can overcome some of these limitations by guiding visitors through the multiple graphics. Still, you will have a difficult time attracting people if you include too many results. A conference attendee strolling by your poster will say to themselves "Wow, that's too much information for me to deal with right now with all of these other posters to look at. I think I'll move on." You want to avoid this reaction.

If you use one of these designs make sure that the relationship between the graphics is clear so that a visitor can easily follow your argument. We will see examples of how to do this when we review some real posters in the next chapter.

Answers to Exercises

Hermaphroditism promotes mate diversity in flowering plants (Christopher *et al.*, 2019)

ANALYSIS

Even for a technical audience, the term hermaphroditism is a bit obscure. A hermaphrodite is an organism that is both male and female. Hermaphroditism is the active noun that is used to describe this type of organism, but this noun is seldom used. Its use here tends to obscure the author's intended meaning. When we remove the term, we get *Plants that contribute both pollen and ovules to the next generation increase the diversity of their mating partners.* Now the title is simple and clear, but not surprising. Naturally, an organism with both types of sex organs will have more mating partners. This title leaves the reader wondering why they should care. The conclusion seems obvious. What is missing is the context of the research. Most previous research on plants with both male and female reproductive parts has not tracked the number of female plants that receive pollen from a single male parent. The authors make this clear in the abstract, but it can also be put in the title. When we include it, we have: *Accurate assessments of reproductive effort in plants must track the diversity of ovules fertilized by a single male.* Now the title is simple, clear and at least somewhat surprising. It tells the reader that there is a problem with the way this type of research has been done, and it tells them how to fix it. The title points to a significant result. The good thing about removing the term hermaphroditism and clarifying the context is that the revised title is also good for a general audience.

Inbreeding reduces long-term growth of Alpine ibex populations (Bozzuto *et al.*, 2019)

ANALYSIS

This is a great title. It is a concise summary of the study's findings. It is very difficult to do better than this. Our one caveat is that the text makes it clear that the authors think that reduced long-term growth would be found in other species if comparable data were available. Their results are therefore an example of a general pattern, not a specific feature of Alpine ibex populations. Given this, we might consider a title like *Inbreeding decreases long-term growth of mammalian populations* or, if the authors wanted to highlight their study species *Inbreeding decreases long-term growth of mammalian populations, evidence from the Alpine ibex.* Both titles make it clear that the authors believe that they have found a general pattern. Both have the potential to increase the number of readers.

Embolism resistance in stems of herbaceous Brassicaceae and Asteraceae is linked to differences in woodiness and precipitation (Dória *et al.*, 2018)

ANALYSIS

This title clearly presents the results of the study by linking embolism to the degree of woodiness and to the amount of precipitation. A slightly better title might replace the word "differences" with a specific term to indicate the types of differences that were found. The new title might look like this: *Embolism resistance in stems of herbaceous Brassicaceae and Asteraceae is linked to increases in woodiness and to precipitation.* If the authors feel that these results are specific to the Brassicaceae and Asteraceae, this title is good. However, if they believe that their results are an example of similar trends in other herbaceous plants, they might want to change the title to *Embolism resistance in herbaceous plants is linked to increases in woodiness and precipitation.* This title can be further simplified by focusing on the most surprising part of the results. It will not be surprising to most plant biologists that plants growing in wetter conditions have lower rates of embolism. It is surprising that the degree of embolism is related to woodiness. A revised title could make this clear: *Embolism resistance in stems of herbaceous plants is linked to increased woodiness.* Or even *Embolism resistance in stems of herbs is linked to increased woodiness.*

Ahlquist, T.K. and Vincent, M.A. (2019) Four new species in the *Trifolium amabile* species complex from the United States, Mexico, and Guatemala. *Botany 2019 Poster: PRT019.*

Barker, N., Roy, C., Cumming, S.G. and Darveau, M. (2012) Predicting waterfowl occurrence and distribution: effects of climate and habitat. Poster presented at the 5th North American Ornithological Conference, Vancouver, British Columbia, Canada. Available at: https://figshare.com/articles/Predicting_waterfowl_occurrence_and_distribution_Effects_of_climate_and_habitat/12404201

Bozzuto, C., Biebach, I., Muff, S., Ives, A.R. and Keller, L.F. (2019) Inbreeding reduces long-term growth of Alpine ibex populations. *Nature Ecology & Evolution* 3, 1359–1364. doi: https://doi.org/10.1038/s41559-019-0968-1

Christopher, D.A., Mitchell, R.J., Trapnell, D.W., Smallwood, P.A., Semski, W.R. and Karron, J.D. (2019) Hermaphroditism promotes mate diversity in flowering plants. *American Journal of Botany* 106, 1131–1136. doi: https://doi.org/10.1002/ajb2.1336

Dahal, S. (2011) Eyes don't lie: understanding users' first impressions on website design using eye tracking. MSc thesis, Missouri University of Science and Technology, Rolla, Missouri. Available at: https://scholarsmine.mst.edu/masters_theses/5128

Dória, L.C., Meijs, C., Podadera, D.S., Del Arco, M., Smets, E., Delzon, S., and Lens, F. (2018) Embolism resistance in stems of herbaceous Brassicaceae and Asteraceae is linked to differences in woodiness and precipitation. *Annals of Botany* 124, 1–14. doi: https://doi.org/10.1093/aob/mcy233

Faulkes, Z. (2021) *Better Posters: Plan, Design, and Present a Better Academic Poster*. Pelagic Publishing, Exeter, UK.

Jantzen, J., Freitas Oliveira, A.L., Soltis, D. and Soltis, P. (2019) Biogeography of *Tibouchina* s.s. (Melastomataceae): identifying the origins of cerrado and campos rupestres plant diversity. *Botany 2019 Poster: PBG008*.

Jiang, L., Hua, Y., Hu, G.-L. and Hua, B.-Z. (2019) Habitat divergence shapes the morphological diversity of larval insects: insights from scorpionflies. *Scientific Reports* 9: 12708. doi: https://doi.org/10.1038/s41598-019-49211-z

Morrison, M. (2021) Better Scientific Poster. Open Science Framework (OSF), Center for Open Science, Charlottesville, Virginia. Available at: https://osf.io/ef53g/

Phillips, H., Bruenn, R., Landis, J. and Specht, C. (2019) Elucidating the roles of MYB-related transcription factors in Zingiberales zygomorphy. *Botany 2019* Poster: PEV003.

Rickart, E.A., Balete, D.S., Timm, R.M., Alviola, P.A., Esselstyn, J.A. and Heaney, L.R. (2019) Two new species of shrew-rats (*Rhynchomys*: Muridae: Rodentia) from Luzon Island, Philippines. *Journal of Mammalogy* 100, 1112–1129. doi: https://doi.org/10.1093/jmammal/gyz066

Stevens, M., Leventhal, L. and Ison, J. (2019) What is the genetic diversity of conspecific pollen carried by native bees to *Echinacea angustifolia*, a prairie perennial? *Botany 2019* Poster: PRP007.

Tian, J., Wang, C., Xia, J., Wu, L., Xu, G., *et al.* (2019) Teosinte ligule allele narrows plant architecture and enhances high-density maize yields. *Science* 365, 658–664. doi: https://doi.org/10.1126/science.aax5482

Tressel, L. and Vincent, M. (2019) Revision of the *Trifolium depauperatum* species complex. *Botany 2019* Poster: PSY022.

Walker, X.J., Baltzer, J.L., Cumming, S.G., Day, N.J., Ebert, C., Goetz, S., Johnstone, J.F., Potter, S., Rogers, B.M., Schuur, E.A.G., Turetsky, M.R., and Mack, M.C. (2019) Increasing wildfires threaten historic carbon sink of boreal forest soils. *Nature* 572, 520–523. doi: https://doi.org/10.1038/s41586-019-1474-y

Analysis of Real Posters

In this chapter we will review some posters and see how their designs meet the three criteria. A poster must attract visitors, serve as the backdrop for your presentation, and must stand on its own and present your results when you are not in attendance. We will be assisted in this task by eye-tracking simulations provided by EyeQuant (EyeQuant, 2009–present). EyeQuant uses advanced neuroscientific research, artificial intelligence, and neural network modeling to simulate the first 3 seconds of how users react to a design. Its clients are major companies like Google, Cannon and Epson who use the technology to produce cleaner and more effective web designs. Though not specifically designed to evaluate posters, the technology can tell us about where a visitor will focus in the first 3 seconds. It will give us a good idea of what a visitor will see as they stroll by your poster.

Poster 1: A Better Poster

The first poster is a classic landscape version of Mike Morrison's Better Poster design (Klöwer *et al.*, 2019). The title, authors names and their affiliations are all in the upper left corner, along with a picture of the first author and his contact information. The title, *Towards 16bit weather and climate models: Posit numbers as an alternative to floating point numbers,* is set at an appropriate font size given its length and supplements the description of the main result, which occupies the center of the poster. If you are going to include both a title and a main finding, we recommend that the main finding elaborates or completes the title as in this example. The authors' names are sized to fit on a single line. We would, however, have liked to have seen them a bit larger as a main take away of the poster should be the name of the first author. The picture of the first author along with his contact information is a nice addition which we strongly recommend. The authors' affiliations are present and are set in an appropriately small font size. The university logo is present at the top of the central section and is not obtrusive. Additional logos are present at the bottom of the left and right columns, in line with

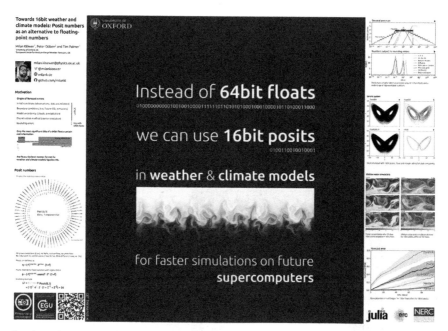

the QR code. This placement makes good use of what otherwise would have been dead space adjacent to the code. Placing supporting logos at the bottom of the right column continues the horizontal line established by the logos on the left and gives a sense of weight and balance to the poster.

The main finding is set in a very large font size at the center of the poster, on a blue background. It is supplemented with an image that has disciplinary meaning. The image shows a simulation of turbulence, which is an important use for supercomputers. The central portion of the poster also contains a very nice visual representation of the difference between a binary representation of a 64-bit floating number, and a 16-bit posit. This representation clarifies the difference between these two somewhat abstract concepts. This is a nice example of how you can represent your findings visually to increase your viewers' understanding.

The left and right columns present the rationale and evidence for the authors' conclusion. Both columns consist primarily of graphics, supplemented with a minimal amount of text. They are intended to be read top to bottom and left to right. The lack of text makes them a bit hard to understand for the non-specialist but makes them ideally suited as backdrops for an oral presentation, which was the main use for this poster. The poster was only on display when the author was present. The columns would be a bit easier to read if the graphics were lined up horizontally across the poster in the same way the logos are at the bottom, but this is a minor point. The presence of the QR code, which links to the full publication, allows interested visitors to gain a deeper understanding of the work. If the poster had been available when the author was not present, the link to the full paper would have offset the difficulty of understanding the graphics. Interested visitors could have scanned the QR code and been directed to the full paper, which includes a detailed description of the graphs.

The EyeQuant eye-tracking simulation shows us why this poster is so effective at attracting visitors. The heat map shows the predicted frequency of eye fixations in the first 3 seconds of a person's gaze. Warmer colors mean more fixations. Cooler colors mean less. Most fixations are on the center of the poster: the central image and the text immediately above and below it. The simulation also shows that viewers have read the main finding, and spent some time looking at the QR code. By focusing on the title and the central image, a visitor immediately knows that the poster is about weather

and climate modeling and that the results are relevant to the modeling of turbulence. These are all key points that will attract attention. In fact, this poster never had less than two visitors at a time for the full 2 hours that it was displayed. At times there was a crowd surrounding the poster.

©EyeQuant, used with permission.

All and all, this is a great poster that does an excellent job of meeting all three requirements. It easily attracts attention, it provides a backdrop for an oral presentation, and it provides a way for interested viewers to find out more. Given the difficulty of meeting these goals, it may seem a bit churlish of us to complain that there is a good deal of blank space at the center of the poster. This blank space is one of the reasons that the poster is so attractive from a distance. It sets off and frames the central image and text but its presence also restricts the size of the marginal graphics, which could have been presented a bit larger. Still, this is a minor issue in an overall excellent poster. The graphics are certainly large enough to see if a viewer is standing immediately in front of the poster, which is the way most visitors would view this poster.

Poster 2: A Classic Three-Column Design

Our next poster uses a classic three-column design with a large central graphic (Tressel and Vincent, 2019). The individual images of the central graphic read as a single figure based on their horizontal alignment and colored connecting lines. This provides a layout that will attract visitors and allows the presentation of detailed results. The text in the center provides just enough information to understand the images. The main text of the poster, which is included because this poster had to stand on its own for most of its display, is in the left and right columns and is presented as bullet points. This makes the information easy to assimilate. Each section of text is

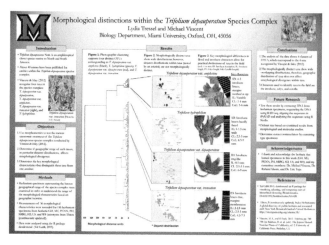

©Lydia Tressel, used with permission.

set off in its own box, which is clearly separated from the other sections by blank space. The background is a double gradient that is darker at the edges and lighter in the middle. This draws the user's attention to the center of the poster where the most important information is presented. Background colors do not often work on posters and, in general, we recommend they should be avoided. However, this is a nice example of their effective use.

Though this is a very good poster, let us see if we can make it even better. The title is a bit hard to see and it does not clearly communicate the major result. It provides a general idea of what the poster is about, but an interested visitor would have to do a good deal of work to find the takeaway message. Although the text boxes and bullet points have been handled very nicely, there is overall a bit too much text. To remedy this, we suggest combining and shortening the introduction and objectives, simplifying the conclusions, removing the section on future research, and reducing the font size in the acknowledgments and references. These changes will help focus the poster on the major conclusions, which are highlighted in the central graphic. Although we have previously argued that methods sections are not important on a poster, this poster must stand on its own without the author present and so there is some justification for including them here. They still must be brief, as they are on this poster.

The logo in the upper left corner and the photograph in the upper right add little content to the poster. The logo actually distracts the user from the central portion of the poster, as we will see when we look at the eye-tracking analysis. If you include a school logo, you should place it in a less prominent position, reduce its size, and perhaps make its background transparent so that the background color of the poster shows through.

The photograph in the upper right corner is very nice and provides an image of a study species, but a very similar photograph appears near the top of the left column where it draws much more of the user's attention. The position of the photo in the upper right corner is intended to balance the school logo in the upper left. However, the difference in color and size of the photograph keeps the poster unbalanced. We would much prefer to see the first author's photograph at the upper right, though with such a strong logo in the upper left even this photograph is unlikely to balance the composition.

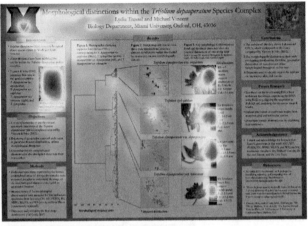

©EyeQuant, used with permission.

The eye-tracking heat map confirms some of the things we have been saying. Visitors look first at the images and at the school logo. It would be better if they looked first at the images in the center. The eye-tracking data indicates that they are looking in the right direction but not at the most informative parts of these graphics. The really bad news is that significant fixation time is spent looking at the school logo and at the photograph in the left column. Neither of these are particularly informative. Also notice that none of the major fixations are on the photograph in the upper right. It is as if this photograph were not there. Keep these things in mind when you design your next poster.

Poster 3: A Variation on the Three-Column Design

The central column of our next poster is split horizontally in order to display two large graphs (Stevens et al., 2019). Drawings of the bees, which are a central part of this research, are placed in each graph. Abbreviations of their names appear on the x-axis. Both elements help to link the graphs to the supplemental information in the right column. The color

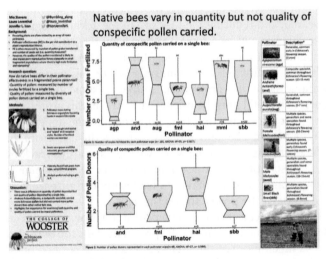

©Mia Stevens, used with permission.

of the graphs is repeated in the background and again in a text box on the left. This common color tells the viewer that these sections belong together. It also suggests that the information in green is subsidiary. This is a very nice way of focusing the viewer's attention on the major findings. The choice of these background colors is particularly good because they work for all types of color blindness. You can check your colors with one of the color-blindness simulators that are available on the internet (Flück, 2006–present). These sites allow you to upload an image and view it through different lenses.

The title on the poster clearly states the main conclusion and invites the viewer to inspect the central graphs for confirming evidence. The authors' names appear in the upper left-hand corner with adjacent contact information. Their affiliations are at the bottom of the same column below the discussion. The text is presented in bullet points and in the figure captions. The college name and a project logo are placed in the lower left corner balanced by a photograph of the authors in the lower right. The citations and acknowledgments are written across this photograph at the bottom of the right-hand column.

There is much to like about this design, especially the central graphs and the clear link between them and the title. The use of background colors to link the results is also excellent, as is the link between the graphical elements and the subsidiary information in the right column. This link would have been even stronger if the same drawings had been used in both the graphs and the right column. The labels on the *x*-axis could also be more clearly related to the species names so that it is easier to connect them to the information on the right. Overall, the text size is adequate, but that set against the green background is a bit small.

The picture of the authors at the bottom right clearly captures their joy while conducting this research but it does not provide a good way of recognizing them when they are not present. This poster was displayed both while the first author was present and for a longer time while she was not. Under these circumstances it is best to include a clear picture of the first author so she can be recognized outside the poster session. Adding a QR code is also a good idea. There is clearly a complex analysis behind these results. Some visitors would undoubtedly like to know more.

At first sight the EyeQuant heat map looks confusing. A visitor's initial gaze does not seem to be drawn to any meaningful area of the poster. The eye-tracking pattern looks like an F. This F-shaped pattern is a

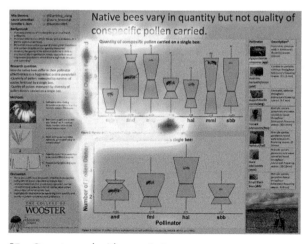

©EyeQuant, used with permission.

common heuristic used to extract meaning from a web page (Nielsen, 2006; Pernice, 2017). The reader starts near the top looking for meaningful content, then scans down as she skims the page. Her eye moves left to right as she reads the text to produce the bars of the F. This viewing pattern fits very well with the layout of this poster. Visitors look at the title, but then their eye is drawn down along the y-axis and then right to the word pollinator on the x-axis of the upper graph. The drawings of the bees are not large enough to attract attention. Though the eye-tracking data show that visitors pay little attention to the central graphs, the data simulates the gaze of an untrained observer. For a scientist, the central graphs will be more meaningful and it is likely that more of their initial gaze will be directed to them. The clearer these graphs, the longer their gaze will linger and the more likely they will be to approach the poster to learn more.

Of the three roles for a poster, this poster does a great job in attracting attention and providing a backdrop for an oral presentation. It is a bit weaker in its presentation of the key results when the author is not present. Simplifying the text, clarifying the focus on the main result, and providing a QR code that links to additional information would address this problem.

Poster 4: Portrait Orientation of a Three-Column Design

Our first poster in portrait orientation also uses the center to good effect (Jantzen *et al.*, 2019). The phylogenetic tree is color coded to match the distribution map of South America so that it is easy to follow the geographic origins of the taxa. The text is placed in color-coded boxes adjacent to supplemental graphics, which are also color coded to the phylogeny. The color coding makes it easy to relate the explanatory text to the phylogenetic tree and distribution map. The text is not excessive, and the methods are presented as bullet points. Although the title does

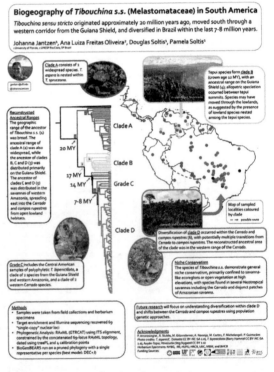

©Johanna Jantzen, used with permission.

not include the main result, it is followed by a helpful summary. The authors' names appear immediately below the summary, in an appropriately large font. The institutional affiliations are small and non-intrusive. There is a photograph of the first author at the top left. The acknowledgments are set in a smaller font at the bottom right of the poster. University logos and other icons also appear in this box.

Overall, this is a very strong design. It would be even stronger with a few relatively minor changes. A better title would clearly state the main result in a large, bold font. The methods, future directions, and acknowledgments could all be moved to a website, linked with a QR code. This would allow a more thorough presentation of the results. The placement of the text boxes and supplemental images is somewhat haphazard. Freeing up some space would allow these to be placed more regularly. This placement might also clarify the sequence in which the boxes should be read.

The EyeQuant analysis shows another F-shaped scanning pattern, though it is weaker than in the last poster. Most fixations are on the top of the phylogenetic tree and on the summary below the title. There are also a few fixations in the region of the green arrow above the map of South America. The good news is that this scanning pattern highlights several important points. The viewer looks at the summary of the findings, looks at the phylogenetic tree, and glances, however briefly, at the map to relate this information to the distribution map.

Overall, this poster does a great job in attracting attention and presenting the main result. It also provides a good backdrop for an oral presentation but is a bit weaker in providing a

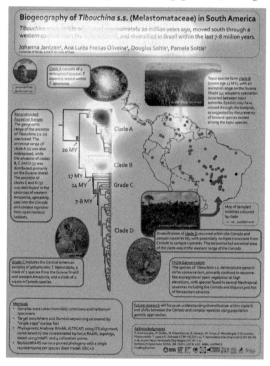

©EyeQuant, used with permission.

visitor with a detailed explanation of the results when the author is not present. Simplifying the arrangement of the text boxes, adding some additional information as bullet points, and clarifying the reading sequence of the boxes would solve these problems. Moving some of the information off the poster and on to a website, linked with a QR code, would make these changes possible.

The designs of our next three posters break up the space into several smaller sections. This allows the presentation of additional details but carries the danger of cluttering the poster and obscuring the central message. Problems with clutter can be alleviated by aligning the presentation boxes with each other and separating them with white space. Boxes that belong together can be abutted.

Poster 5: Block Design 1

The poster by Phillips *et al.* (2019) nicely demonstrates these features. The poster title is centered at the top in a large font size. The authors' names appear directly below, with footnotes linking to their affiliations. An introduction is immediately below their names. The school logos are in the upper left-hand corner, balanced by the QR code that links to detailed information about the methods, in the upper right. Abbreviated methods are in the two boxes at the bottom of the poster. There are no acknowledgments or references. All the boxes contain graphics, supplemented with text. The graphics are large and are attractively arranged. The five horizontal bands are given slightly different colors and are separated by white space so that it is easy to see how they are thematically linked. The graduated green background at the bottom gives the poster an attractive substantiality. The poster is beautifully executed.

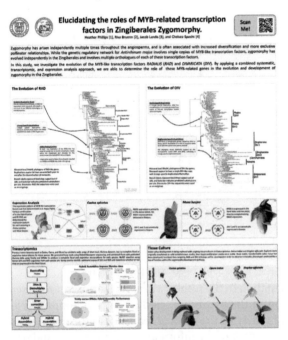

©Heather Phyllips, used with permission.

The problem with the poster is that it does not clearly present the authors' main result. There are several results presented in the upper boxes, but these have not been formed into a single coherent message. Each box stands on its own without a clear relationship to the others. The title and adjacent text introduce the research, but do not present the main result. The information in the two methods boxes at the bottom does not contribute significantly to the overall meaning and could be removed.

Although we prefer to see text presented as bullet points, that would be difficult in this poster, especially in the area immediately under the title. Of course, a photograph of the first author should always be included.

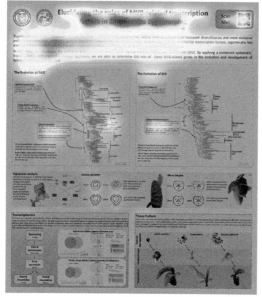

©EyeQuant, used with permission.

The eye-tracking data shows that virtually all the initial fixations are on text, not images. Viewers are looking primarily at the title, with only slightly less attention given to the introduction. The QR code also gets a significant amount of attention. Only a few places in the main body receive any fixations. Because the phylogenetic trees are in black and the text is in a small font size, they receive relatively little attention. With this design visitors are looking primarily at the title and the top of the poster.

Of the three crucial functions for a poster, this one is good at attracting attention. It is attractive and beautifully laid out. It also provides a good backdrop for an oral presentation but is weaker in presenting the main result when the author is not present.

Poster 6: Block Design 2

Taxonomic research, especially the description of new species, presents its own set of problems. The main results of these studies are species descriptions, which do not translate well to posters. The poster by Ahlquist and Vincent (2019) is an exception. Both the layout and the title of the poster clearly show the main result, the recognition of four new species of clover.

The poster is laid out symmetrically with the center occupied by four distribution maps. These maps provide a nice focus at the center. The main text is above them and is fully justified. This supports the symmetrical layout of the poster. Drawings of each species are arranged around the central maps to make good use of the somewhat unusually shaped spaces left by the maps and title. The distinctive characteristics of each species are indicated in green on each drawing. The green used for this text is easily visible even to people who are green blind (deuteranopia). The main body of the poster is framed by the four new species names, set vertically. These names, along with the logos and text at the bottom, give the poster a sense of solidity. The acknowledgments and references are set in an appropriately small font size at the bottom of the poster.

The most impor-
tant part of the title
is simply stated and
set in a larger font
size. Its position at
the top and center
of the poster, beau-
tifully flanked by
illustrations, leaves
no doubt about the
takeaway message.
The authors' names
and affiliations are
immediately under
the title, in columns
placed symmetrically
to reflect the sym-
metrical design of
the poster. The univer-
sity logos are placed
symmetrically and

©Tia Ahlquist, used with permission.

unobtrusively at the bottom of the poster. The fact that they are red makes a nice
contrast to the green that is used in the rest of the poster. This contrast works best for
people with normal color vision. Viewers with red–green color blindness may have
problems distinguishing red from green. However, use of a color-blind checker

(Flück, 2006–present)
shows that the red and
green used here can be
easily distinguished even
for people who are red
blind (protanopia). Still,
it is best to avoid using
these colors together.

The EyeQuant data
shows two major regions
of fixation: (i) imme-
diately below the title;
and (ii) the upper left
center. This is good news
as these regions are rel-
evant to the main point
of the poster. The focus
on the upper left of the
four images is likely an
outcome of an F-shaped

©EyeQuant, used with permission.

scanning pattern. Visitors initially scan left to right at the top, and then scan downwards looking for additional relevant information.

This poster is so nicely and symmetrically designed that we are hesitant to suggest changes. However, we feel it would be slightly better if the green text had been a bit darker, and the dots on the distribution maps had been in color. It would also have been nice if the distribution maps were identical, so that it would be easier to compare the distributions of these species. Of course, we always like to see a picture of the first author, and it would be good if her contact information had been provided. A QR code linking to the full species descriptions would also have been a nice touch.

This poster does a very good job of meeting all three functions for a poster. It is attractively designed and will be sure to attract visitors. The drawings are well executed by the first author and provide an excellent backdrop for her presentation. The green text that indicates the defining characters of each species makes the poster easy to understand when the author is not present. Overall, this is a great design. A QR code with a link to the full species descriptions, and a picture of the first author would only make it better.

Poster 7: Block Design 3

The last poster we will consider uses an unconventional and innovative design (Barker *et al.*, 2012). We have included it to show one way that you can present a good deal of data while maintaining audience interest. The poster is framed by four images, one of which is of the first author. These images give the poster a sense of stability that would otherwise be lacking. The use of images in circles in an arc at the center of the poster tends to destabilize the design. The corner images counter this instability. The title is set in an appropriate

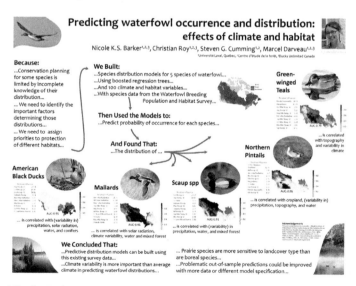

Nicole Barker, CC-BY 4.0[1]

font size that will be easy to see from a distance. It is right justified, which draws attention to the photo of the first author. The authors' names appear immediately below the title with their affiliations under their names in a much smaller font size. The title box is balanced at the bottom of the poster by the conclusions. The positions and sizes of these two areas also helps stabilize the design. The center of the poster is used for a series of boxes with photographs and distribution maps. Each distribution map is accompanied by a table listing the key variables used in predicting that distribution. A short description of the most important variables is given below each map. Arrows are used to indicate the reading sequence of the maps. All the text, except the acknowledgments, is presented as bullet points. The acknowledgments are set in a much smaller font size in a section left blank by the design.

There is much to like about this design. The distribution maps and accompanying tables present a great deal of data. The photographs help the viewer link the ducks to their distributions. The arrows provide a clear reading sequence. The text is unobtrusive and informative. The inclusion of a QR code could have provided a link to more detailed data, or a copy of a published manuscript. However, this poster was presented in 2012, which would have been one of the earliest uses of QR codes on a poster.

Our main concern with this poster is that the central point has gotten lost in the data. This is a frequent problem when a poster presents a lot of data. It is difficult to balance this presentation with a clear message. In this case the main message seems to be that it is possible to construct predictive models of waterfowl distribution. This is an important message for anyone working on environmental models, but could the message be more broadly focused? The importance of climate variability is mentioned as one of the points in the conclusion. This could be emphasized in the title. A revised title might read _Climate variability is more important than average climate in predicting waterfowl distribution._ This title still conveys the role of modeling but focuses on one of the important conclusions from these models.

The eye-tracking data shows that visitors are looking at very relevant parts of the poster. The highest number of fixations occurs at the center, over the first part of the

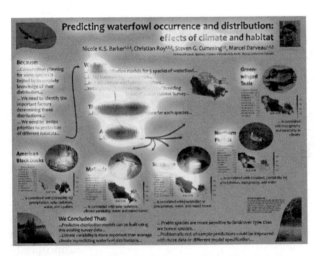

©EyeQuant, used with permission.

results. A high number of fixations also occurs over the explanation of the model ("We Built"), with slightly lower rates over the background information ("Because") and the species names. Some attention is also given to the title. We really could not hope for a better result given this design. In the first 3 seconds a visitor will have learned that the authors built species distribution models and that these models predict the distributions of five species. Clarifying the title and setting the conclusions in a slightly larger font size would only enhance this understanding.

All three functions of a poster are handled well by this design. It will attract visitors, provide a good backdrop for a presentation, and is easy to understand when the author is not present. Changing the title to focus on the core finding would only increase its quality.

Ahlquist, T.K. and Vincent, M.A. (2019) Four new species in the *Trifolium amabile* species complex from the United States, Mexico, and Guatemala. *Botany 2019 Poster: PRT019.*

Barker, N., Roy, C., Cumming, S.G. and Darveau, M. (2012) Predicting waterfowl occurrence and distribution: effects of climate and habitat. 5th North American Ornithological Conference, Vancouver, British Columbia, Canada. Available at: https://figshare.com/articles/Predicting_waterfowl_occurrence_and_distribution_Effects_of_climate_and_habitat/12404201

EyeQuant (2009–present) EyeQuant. Available at: https://www.eyequant.com/

Flück, D. (2006–present) Colblindor. Available at: https://www.color-blindness.com/

Jantzen, J., Freitas Oliveira, A.L., Soltis, D. and Soltis, P. (2019) Biogeography of *Tibouchina* s.s. (Melastomataceae): identifying the origins of cerrado and campos rupestres plant diversity. *Botany 2019 Poster: PBG008.*

Klöwer, M., Düben, P.D. and Palmer, T. (2019) Towards 16bit weather and climate models: Posit numbers as an alternative to floats. Poster presented at conference European Geoscience Union (EGU) 2019, Vienna, Austria. Available at: https://doi.org/10.13140/RG.2.2.20921.01128

Nielsen, J. (2006) *F-Shaped Pattern For Reading Web Content.* Nielsen Norman Group. Available at: https://www.nngroup.com/articles/f-shaped-pattern-reading-web-content-discovered/

Pernice, K. (2017) *F-Shaped Pattern of Reading on the Web: Misunderstood, But Still Relevant (Even on Mobile).* Nielsen Norman Group. Available at: https://www.nngroup.com/articles/f-shaped-pattern-reading-web-content/

Phillips, H., Bruenn, R., Landis, J. and Specht, C. (2019) Elucidating the roles of MYB-related transcription factors in Zingiberales zygomorphy. *Botany 2019 Poster: PEV003.*

Stevens, M., Leventhal, L. and Ison, J. (2019) What is the genetic diversity of conspecific pollen carried by native bees to *Echinacea angustifolia*, a prairie perennial? *Botany 2019 Poster: PRP007.*

Tressel, L. and Vincent, M. (2019) Revision of the *Trifolium depauperatum* species complex. *Botany 2019 Poster: PSY022.*

Audience—It Is All About the Audience

8

The most important thing to consider when planning your presentation is your audience. Your audience will determine how much technical language you can use, how much background information you must present, and even what clothes you wear. As scientists we do not often consider our audience when we plan our presentations. We want our data to speak for itself and, since data is independent of the audience, we think that its presentation will be sufficient. We spend so much time collecting and verifying the data that it is inconceivable that others will not understand them as intuitively as we do. We forget that in collecting the data we have gone through a long process; a process that is fraught with trials and setbacks. Presenting only the data short-circuits this process. Scientists who work in the same area will, of course, want to see your data. They will know the techniques and have the necessary background to understand it. Everyone else will be lost. They will find the data hard to understand because they will not have a deep understanding of the techniques that generated it. This is one reason so many speaking coaches say that your presentation is not about you, it is about your audience (Craig and Yewman, 2014). As Bennett (2013, p. 25) says, a speaker should not ask "What do I want to share?" but rather, "What does the audience want to hear?" We might even say, "What does the audience *need* to hear?"

Types of Audiences

In Chapter 4 we introduced three different audiences: (i) a technical audience of scientists who work in your immediate field (ii) a professional audience of scientists who work outside your field; and (iii) the general public. These audiences will all have different expectations about your presentation. Let us look at how you can address these audiences effectively.

A technical audience of scientists who work in your immediate field will know all the jargon and most of the background necessary to understand your presentation. For this group you can use as much jargon as you like.

You may even be judged negatively if you do not use jargon. Everything in your training has prepared you for this audience. You know the literature, the jargon, the necessary background. These scientists will understand the significance of your work and will be interested in it because it relates to their own. They will be eager to learn new results.

For these presentations it is a good idea to think in advance about who is likely to be at your talk. If you expect someone whose work has had a significant impact on yours to be present, it might be a good idea to introduce yourself to them prior to the start of your talk. By connecting with an influential scientist you will be able to find a friendly face in the audience. This will increase your confidence and improve your audience connection.

Let us consider a fictitious example. Let us say you are presenting a paper on tropical lowland ecology at a tropical ecology symposium in Costa Rica. Given the nature of the meeting you have every reason to expect that Dr. Wagner, one of the most highly respected authors in your field, will be in the audience. Of course, if his work bears directly on yours, you will reference it in your talk. However, your work may not be directly related to Wagner's. Still, there he is in the audience, one of the leaders in the field. You can introduce yourself to him prior to your talk and recognize his presence with a brief statement along these lines: "It is a pleasure to be here today in the presence of scientists like Dr.

Wagner whose work on tropical ecology has shaped so much of the field." An introduction like this, with a brief nod in Dr. Wagner's direction will provide a better connection with the audience than 15 minutes of platitudes. You will find that the audience is much more attentive to your work, and much more likely to remember your name.

Now let us suppose that you are speaking to a larger, more general technical audience at a society meeting such as the Ecological Society of America or Australia, the British Ecological Society, or the European Ecological

Federation. These scientists will be familiar with your general area but will not be as cognizant of the details as will scientists who work in your immediate field. Although this may seem to be a very different venue, the need to connect with the audience is the same. In fact, it may be even more important to connect with this audience as the potential impact of your talk is greater. To address this need, Alan Alda, the guru of scientific communication, suggests catching the eye of an audience member:

> I know how important it is to look into the eyes of an audience when I'm giving a talk. I don't just scan the audience; I catch the eye of individual people and hold their gaze for a few seconds. When I do that, something happens between us. I'm actually talking to someone, not just saying the words I've prepared, and as a result something changes in my tone of voice. It becomes more personal and direct. And I get reinforcement from the warmth I see in their faces.

> (Alda, 2017, pp. 120–121)

This feeling of warmth is something that storytellers know very well, and the best scientific communicators are great storytellers.

In some ways these professional audiences are the most difficult. You will need to tell them why your work is important and what it means. You must explain many aspects of your work but at the same time you must not talk down to them. They will expect you to use technical terms even if they are new to them. They will want to be treated like the intelligent people that they are. This means that you must explain the technical terms in a nonthreatening way. You are explaining them to intelligent people, who just happen to work in a different field.

Let us say you want to speak about the part of a plant's life cycle that produces the sex cells, the gametes. This part of the life cycle is called the gametophyte, the gamete-bearing plant. To introduce this concept you might say, "Unlike animals, plants have a separate part of their life cycle that produces the gametes. This part of the life cycle is called the gametophyte. The term literally means "gamete plant." It is the part of the life cycle that bears the gametes." This is a much better way of proceeding than to say, "This is the gametophyte, the part of the life cycle that produces the gametes." While the information content of these two approaches is the same, they say very different things about your attitude toward the audience. The first emphasizes their intelligence. The second emphasizes yours. The first says, "You are smart enough to understand this. Now you know a term that helps you express this understanding." The second says, "Here is a new term. I will now explain it to you." You will find your audience much more receptive if you first address their intelligence and then give them the terms with which to express their new understanding.

Although speaking to the public may seem more difficult, in some ways it is the easiest type of presentation. It is easy because the audience will not understand your field and so will not be critical of small errors. Of course, you will have to explain your work in non-technical terms, and you will not be able to use jargon. If you find that you must use technical terms, you can give the definition first, then introduce the term as we just described. You should also use similes and metaphors to explain complex ideas as we discussed in Chapter 3. Take the time to learn as much about your audience as you can so that you can choose metaphors that will be meaningful to them.

Despite their differences, there is one important similarity between these three audiences: your connection with the audience. Scientists in your field will understand the significance of your work and will be interested in it because it relates to their own. A general audience will not share this interest. You will have to convince them. If, like most scientists,

- banana evolution -

you work on something that is not in the news, you will have to find a way to convince them to listen to you. The best and most effective way to do this is to speak with enthusiasm (see Chapter 9). By being enthusiastic you will engage the audience and convey a feeling of excitement about your work. Even if you work on bananas as does the author, you cannot assume that the audience will know or care about your research. They will know what a banana is, but they will not know about banana evolution. They will learn to care about these things because you care about them. They will learn that you care because you are engaged and connect with them with enthusiasm.

Distracted

Bored

Engaged

Creepy

Disengaged

If you stand with your back to the audience speaking in a monotone, the content of your presentation will get lost as the audience focuses on your poor skills. It is hard to pay attention to what someone is saying if they appear disengaged or uninterested. This type of presentation will damage your reputation, perhaps beyond repair. As Professor of Zoology Robert Anholt says, "The speaker's attitude with respect to the audience often determines whether the presentation will be clouded by an atmosphere of skepticism or received in a welcoming ambience of motivated interest" (Anholt, 2006, p. 2).

Here are two ways you can think about presenting to the general public. Steve Curwood, the producer of the US Public Radio program *Living on Earth* (Curwood, 1991–present), envisions his audience as a curious middle-school student who is eager to learn about science (Dean, 2009). Vernon Booth, the author of *Communicating in Science*, suggests that you speak in a way that allows you to be easily understood by someone whose first language is not English (Booth, 1993). These ideas can be touchstones for you as you prepare your public presentations.

Persuasive Presentations

When giving a talk intended to persuade it is imperative that you learn about the nature of your audience at a deeper level than in a normal scientific talk. Some years ago I learned how necessary this is. It was during a talk on fruit bat ecology at the Royal Botanic Garden, Sydney. Here is my recollection of the talk.

> During my time at the garden there was a large colony of fruit bats roosting in the trees each night. The bats were rare and decreasing in numbers, but their colony was damaging valuable trees. There were ongoing discussions about what might be done to decrease the damage, but no action was planned. The speaker was a graduate student who was studying the bats. There was great interest in her talk among the garden staff. Unfortunately, she made two mistakes that decreased the impact of her seminar. First, she adopted an adversarial attitude toward the audience. She assumed that the staff would act to preserve the trees to the detriment of the bats. This was not true. The staff was very concerned about the bats. Her attitude came out in her stance, her tone of voice, and in the message she delivered. Compounding this problem, she failed to do her homework and know who the decision makers were. The director of the garden was sitting in the front row but the speaker did not know this. During the question period she reacted to one of the staff scientists as if he were a primary decision maker. In fact, he had no direct role in any decisions about the bats. She was looking for someone to blame for the harm she thought would be done. When one of the staff scientists asked a question, she responded to him with an inappropriate level of hostility. After her talk I pointed out that the director of the garden was in the front row and that the person to whom she responded had no role in the decision-making process. She seemed generally shocked. She was responding out of her own preconceived ideas and had not taken the time to learn about her audience.

Do not make the same mistake. If you are giving a talk designed to persuade you need to know your audience. You need to know as much as you can about the people to whom you will be speaking, to their roles, and to the decision-making process. Armed with this knowledge you can address their concerns and advance your cause.

If you are speaking to a policy maker, you need to relate your message to issues they care about. These issues often concern the economy, competitiveness, jobs, etc. Do your research and find out what is important to them. Pitch your presentation to these interests. Policy makers do not have time to learn about the details of your work. They want you to get right to the point. Tell them why they should care. You need to know about their politics, about what is important in their district, about what they care about. Your message needs to be simple, focused, and short (Dean, 2009).

When speaking to a journalist, keep in mind that they are looking for a story with an emotional, human element. They want a human story because it will attract attention. They want to tell a story that is appealing to their audience. Your message will be received better if you take the time to learn about who reads their paper, listens to their podcast, or watches their program. Remember that when you speak to them, you are speaking to their audience. If you are speaking to a broadcast journalist and they are going to give you airtime, you must present your points in 2 minutes or less. Two minutes would be a very long time for a program to give you. Use the techniques you have learned in this book and your presentation will have a substantial impact.

Alda, A. (2017) *If I Understood You, Would I Have This Look on My Face?* Random House, New York.

Anholt, R.R.H. (2006) *Dazzle 'em with Style: the Art of Oral Scientific Communication*, 2nd edn. Academic Press, Burlington, Massachussets.

Bennett, T. (2013) *The Power of Storytelling*. Sound Concepts, American Fork, Utah.

Booth, V. (1993) *Communicating in Science*, 2nd edn. Cambridge University Press, Cambridge, UK, p. 78.

Craig, A. and Yewman, D. (2014) *Weekend Language: Presenting with More Stories and Less PowerPoint*. DASH Consulting, Portland, Oregon.

Curwood, S. (1991–present) *Living on Earth®*. Public Radio's Environmental News Magazine, World Media Foundation, Cambridge, Massachusetts. Available at: https://www.loe.org/

Dean, C. (2009) *Am I Making Myself Clear?* Harvard University Press, Cambridge, Massachusetts.

Presentation Skills

9

If someone were to ask us for one tip that would help them become a better speaker, we would tell them to speak with enthusiasm. To speak with enthusiasm means to feel excited and happy and to express these feelings through your gestures, intonation, and energy level. These behaviors will help you connect with your audience. With a good connection your message will be better received and understood.

What do you need to do to speak with enthusiasm? At the very least you would need to:

1. Face the audience.
2. Use expressive gestures.
3. Speak loudly and with varied intonation.
4. Maintain a high energy level.
5. Express interest in your research and show that interest to the audience.

If scientists are perceived as aloof, intellectual, and disconnected from the concerns of life it is most likely because that is how we present ourselves. Our presentations are usually so focused on data that we forget how much depends on our presentation skills. With good presentation skills you can convince even the most die-hard skeptic. Let me give you a personal example.

> Several years ago, I was invited to give a talk at a public library in a very conservative part of my state. The talk was not on anything controversial, but during the question period one of the audience members asked my opinion of global warming. The previous hour had been spent talking about how sexual reproduction takes place in plants. The audience members dissected lily flowers and learned their parts. They learned about the biggest and smallest flowers in the world and about false flowers. They learned about orchid pollination and about unisexual flowers. All this was done while maintaining a good connection with the audience. This connection was likely the reason I was asked about global warming. An audience member felt comfortable enough to ask a controversial question. In answering I did not cite facts and figures. I did not give any data. What I did was to tell a brief story about how

when I first arrived at my university in the mid-1980s we were teaching about the effects of increasing CO_2 and what would happen if the trend continued. I said that it seemed to me that the predictions from that time were now coming true. When I finished the audience member who asked the question said "Wait a minute. I think I am changing my opinion." I believe that this openness to seeing things differently occurred because of my connection with the audience. The audience member felt comfortable enough with me to listen to my answer and adjust his opinion accordingly.

When an audience member is willing to see things differently you know you have made an impact.

Despite what many of us think, scientific talks are performances. Think about any of the great scientific talks you have heard. Would you have found them so compelling if they had been delivered in a deadpan manner? Probably not. A deadpan delivery forces the audience to focus on the con-

tent of the talk. While the content is important, your delivery is of equal importance. The audience must accept and relate to you if you want them to understand and remember your talk. In some cases your career will depend on this. In the end, it is more important that they remember you than your research. You want to be invited to give talks, sit on panels, and join research teams. To achieve this, you need to develop stage presence.

Stage presence occurs when you are so familiar with your presentation that it flows comfortably. You live completely in the moment, enthusiastically conveying your content and connecting with the audience. You have mastered the content and the technical speaking skills that make this possible.

To develop stage presence, you must know your material well enough to dispense with notes. That means, if you use slides, they should contain little or no text. This will remove any temptation to repeat the information on the slide or, worse yet, to read the text. Reading your slide puts your focus on the slide not the audience. It is a sure way to break your connection. We have all been at talks where the speaker reads directly off his slide. We know how annoying this can be.

Body stance, gestures, and facial expression all contribute to stage presence. You should station yourself in front of your audience, not behind a lectern. Your posture should command attention even if you are seated. If standing, you should have a confident stance, with your feet shoulder-width apart pointing forward; knees slightly bent, never locked. This will create balance and provide a solid foundation for your talk.

Your gestures should align with the emotion of your speech. Use them to express your excitement for your research. Keep them open and directed toward the audience as does Lydia Miller in her talk linked later in this chapter (Miller, 2018). Your hands should be open, not in

fists, never in your pockets. If you need to gesture at the audience, use your whole hand with an open palm. Do not point, as this can appear aggressive. If you want to indicate the whole audience, you can sweep your open hand across the auditorium with your fingers extended outwards.

Your facial expressions should also enhance your message. When your face matches your message, you appear more authentic and your credibility increases. Smile and nod your head when appropriate. The audience should never doubt that they are the focus of your attention.

It is better if you do not look down. If you must look at written notes, your glance should be brief. Watch the video of US President Ford's speech at Tulane, linked below, for an example of how looking down can disrupt your talk (Ford, 1974). If you need notes you can use the presenter mode in your presentation software, print your notes in a large font on paper, or display them on a hand-held tablet. With your notes in a large font you can hold them in your hand and quickly glance at them to refresh your memory. If you need extensive notes or must refer to them frequently, you are not well enough prepared for your talk. The audience needs to know that you care enough about them to have taken the time to prepare.

Examples

How do great speakers connect with their audience? Our primary examples will be from speeches by prominent US politicians. We use them because they show a range of speaking abilities and because the videos are in the public domain. We are not concerned with the content of these talks. We focus instead on the speakers' presentation skills and connection with their audience.

In 1988, then US President Ronald Reagan gave a speech to the students at Moscow State University (Reagan, 1988). We will look at how he maintains a connection with his audience under somewhat difficult and unforgiving circumstances. Watch the first minute of the video to see the venue.

Moscow State Venue

The auditorium where Reagan spoke was large and imposing. It was a big, forbidding space in the grand Soviet style. A lesser speaker would have been intimidated by the setting. Reagan, however, maintains a connection with the audience with a minimal use of gestures, which anyway would not have been visible in the space.

Reagan looks from side to side

During the main part of the speech, he continually looks from side to side so that he includes every part of the audience. Watch 30 seconds of the second video to see how he does this. Although he has a teleprompter, he never turns his full attention on it. He glances at it in passing. His attention is always on the audience. It is his main focus.

Question period joke

During the question period he is more animated. When a joke falls flat, he smiles and chides the audience with "You're great. ... Carry on." He does not take the failure personally. He does not get upset. If anything, he has a better connection with the audience after the joke fails. His body language is more open and he is relaxed.

He will later lose much of this connection when he is unprepared for a question. We cannot, therefore, attribute his ease after the joke to the fact that it is the question period and not the formal part of his speech. When he is forced to focus on his answer in response to a difficult question his gestures are less controlled, he glances down more often, and his verbal delivery is less even. He even appears slightly nervous, or even insecure. The contrast between these two incidents shows the importance of practice. The joke was clearly something that Reagan had used in other talks. He was prepared for a good reaction but was not upset when he did not receive one. When he is faced with a question for which he is unprepared he loses focus and his connection with

Question period answer

the audience suffers. He must think carefully about his answer. This throws him off. Even an excellent speaker who is not prepared can lose his connection and fail to communicate effectively.

US President Richard Nixon also developed a good connection with his audience when he spoke to students at the Oxford Union in 1978, 4 years after his resignation due to the scandal of Watergate (Nixon, 1978). Nixon was speaking without notes in response to questions from the audience. Like Reagan he frequently glances from side to side and, in this more intimate setting, he uses many gestures. By the linked point in his speech he is realizing that he has connected with the audience and both sides are responding well. The result is a warmth of presentation that we rarely saw from him when he was president. He appears presidential and statesmanlike, a persona he worked hard to project after he left office.

Nixon at the Oxford Union

Contrast these talks with the short clip from a speech by former US President Gerald Ford in an address to the graduating class of Tulane University in 1974 (Ford, 1974). Do you feel that he is reaching the audience? Although he glances up at all parts of the audience, he also frequently reads from his notes. This is before the advent of the teleprompter, so he had no choice but to use notes. However, he glances at them so often that he distances himself from the audience. If you listen carefully you may also notice that the strength of his voice frequently increases when he is reading and decreases when he looks up. This also breaks his connection with the audience. One is left with the impression that he cares more about the words he is speaking than the audience. The words are certainly important, but to be effective they must impact the audience. This impact only comes through a connection with the speaker. President Ford was never known as a compelling speaker. His performance in this video demonstrates why.

Ford at Tulane

Lydia Miller, Executive Arts Director of Aboriginal and Torres Strait Islander Arts at Australia's national arts funding body, the Australia Council, demonstrates many of the behaviors we advocate in her talk on cultural resilience (Miller, 2018). She looks at the audience, smiles, and uses good gestures. You can see her enthusiasm as she seems to become bigger and more alive. These behaviors improve the delivery of her message. They will do the same for you.

Lydia Miller on resilience

How is it that some of these speakers connect well with their audience, while others do not? The frequency with which they look at the audience is, of course, a major factor. Every time President Ford looks down, he breaks his connection with the audience. The continual focus of Presidents Reagan and Nixon on the audience creates a much better connection.

Another major factor in good audience connection is the cadence of your speech. Cadence refers to the flow or rhythm of spoken words. The cadence is governed by the use of pauses. Let us look at the cadence of a few sentences of Reagan's and Ford's speeches. As we do this, we will also consider which words are emphasized by these two speakers. We will use **bold** text to indicate words or phrases that are spoken with greater emphasis. These changes in emphasis are very apparent in President Ford's speech, but less so in Reagan's. Reagan uses pauses much more than changes in emphasis to punctuate his speech. There are changes in emphasis, but compared to Ford's they are very minor and so we have not annotated them in Reagan's transcript.

Reagan at Moscow State

But freedom doesn't begin or end with elections [pause]. Go to any American town [short pause], to take just an example [short pause], and you'll see dozens of churches [short pause], representing many different beliefs [short pause]—in many places, synagogues and mosques [short pause]—and you'll see families of every conceivable nationality [short pause] worshiping together. [pause]

Reagan at
Moscow State

Go into any schoolroom [pause], and there you will see children being taught the Declaration of Independence [short pause], that they are endowed by their Creator with certain unalienable rights [short pause]—among them life, liberty, and the pursuit of happiness [short pause]—that no government [very short pause] can justly deny [pause].

(Reagan, 1988)

Ford at Tulane University

As we strive together [pause] to perfect a new agenda [pause], I put high on the list of important points the maintenance of **alliances and partnerships** [short pause] **with other people** and other nations [pause]. These **do** provide a **basis of shared values, even** [pause] as we **stand up** with determination for what we believe. [pause]

This, of course, requires a continuing commitment [pause] to peace [pause] **and a determination to use our** good offices [pause] wherever possible [pause] to promote better relations between [short pause] nations of this world.

Ford at Tulane.

(Ford, 1974)

Reagan uses pauses to break his speech into short sections following the punctuation in the transcript. This gives his speech a natural flow that is easy to follow. Ford's pauses are more random. They coincide with his glances at his notes, not with the punctuation. His speech does not have a natural cadence. It is harder to understand. We see an example of this in his final sentence where he pauses after the word commitment in addition to after the word peace. This leaves the word commitment with no clear referent and associates the word peace with determination. The sentence reads as if he wanted to say: a continuing commitment to peace and determination. However, the transcript shows that he meant to say we have a "continuing commitment to peace, and a determination to ..." His placement of the pause obscures his meaning. If we wanted to capture the intended meaning, we might place pauses like this: "This, of course, requires a continuing commitment to peace [pause] and a determination to use our good offices [short pause] wherever possible [short pause] to promote better relations between nations of this world." Read this section of his speech out loud both ways to see the difference.

President Ford's use of emphasis also disrupts the flow of his speech. In the same sentence we are considering, he emphasizes the words "**peace and a determination to use our.**" There are several problems with this emphasis. First, this is not a natural phrase that conveys a clear meaning. The emphasis leaves the audience wondering what it is that will be used. When they hear the next words, "our good offices," uttered without emphasis we can only wonder whether President Ford believes that the USA has good offices or not. A more natural emphasis would be to emphasize the phrase "**a determination to use our good offices**" or even just the word "**determination.**" This would clearly signal that the USA was determined to use its power for good. Read the sentence out loud with the following different emphases to see how they clarify, obscure, or change the meaning.

> "This, of course, requires a continuing commitment to peace and **a determination to use our good offices** wherever possible to promote better relations between nations of this world."

> "This, of course, requires a continuing commitment **to peace** and a determination to use our good offices wherever possible to promote better relations between nations of this world."

> "This, of course, requires a continuing commitment **to peace** and a determination to use our good offices **wherever possible** to promote better relations between nations of this world."

> "This, of course, requires a continuing commitment to **peace and a determination to use our** good offices wherever possible to promote better relations between nations of this world."

As scientists, we do not often think about how our cadence and emphases affect our meaning. However, when speaking to any but the narrowest technical audiences these aspects will be as important than the content. Even scientific audiences outside of your discipline will look to your patterns of speech to help them understand your meaning. Your clear use of pauses and emphases will help them remember your talk much better than if you focus on the technical details, many of which will go over their heads.

Your Turn

There are some excellent talks on the internet that show the speaker connecting well with the audience (Robinson, 2006; Sinek, 2009; Johnson, 2013). An internet search will turn up some to watch. Jump to the middle and watch a few minutes. Look at the speaker's body language and use of gestures. Listen to how they use pauses and emphasize certain words. For this exercise, we are not concerned about the content of the talks.

In the following sections we suggest some activities to improve your presentation skills. Start by practicing them on your own, then try them with a partner. Work on the skills that make the most sense to you. Develop your own personal delivery style, one that feels natural and comfortable. If you have an elevator pitch, you can use it to practice. You can record your practice sessions so you can review them later. If you teach, you can try your skills during one of your lectures. Look for ways to communicate enthusiasm as you speak. Your connection with your students will improve. They will see you as more approachable. Finally, when you are ready, try your skills in front of a technical audience. Start by inviting your lab mates to sit in. Perhaps your department would be interested in hosting a friendly elevator pitch competition. The ones we have seen have been very well received by both faculty and students.

Body and stance

If you sit while speaking, skip the next paragraph.

Stand and look down at your feet. They should be shoulder-width apart. Be sure your feet are pointed forward, toward where the audience will be. Bend your knees slightly. You should feel stable and relaxed. Practice this stance until it becomes

second nature. Your goal is to create a memory of how it feels to have a balanced stance so that you naturally fall into this stance during your presentations.

Whether sitting or standing, work on your posture. If you are able, straighten your spine and push the top of your head up as high as it will go. Imagine you are a marionette with the string attached to your head pulling you upright. This will pull your shoulders back and your chest up and forward. You will feel taller and more confident. Expanding your chest like this improves your blood flow, enhances your balance, and improves your lung capacity. Try this as you move, when you are sitting, or in your next online meeting. People will notice.

Head and expression

Try nodding your head slightly up and down as you speak. Do this as you make important points about your research. This will signal your interest in what you are saying. Now add a smile here and there. You do not need to smile the whole time, only when you are feeling especially good about what you are saying. Avoid a smile that comes off as a grimace. When you work on eye contact, avoid an intense gaze that might be perceived as staring. You want to give the impression of interest, not of a threat. Go back and review Alan Alda's advice about looking at the audience in Chapter 8.

Speech

Start with a very short text from a fairy tale or story and practice reciting it with different emphases. Exaggerate the pauses and changes in intonation so that the text takes on different meanings. You can practice with the text from President Ford's speech reproduced above, try the text we have provided below, or find your own. Here are the opening lines from the Grimm's Fairy Tale *Rapunzel*. After all, Rapunzel lettuce is *Valerianella locusta* (Caprifoliaceae), and a good starting point for a scientific story. Here are two variations you can try. **Bold** text indicates words or phrases that should be spoken with greater emphasis.

> Once upon a time [pause] in a **faraway land** there lived a cobbler and his wife [short pause]. They **desperately** wanted to have children [pause], but **however they tried** [short pause], it was all in vain.

> **Once upon a time** in a faraway land [pause] there lived a **cobbler and his wife** [short pause]. They desperately **wanted to have children** [pause], but however they tried [short pause], it was **all in vain**.

Of course, you will never be this demonstrative in a technical talk, but by practicing these exaggerated exercises you will get used to using pauses and changes in intonation and will learn their power.

Ford, G. (1974) *Remarks at Tulane University*. Presidential Speeches, University of Virginia, Miller Center. Available at: https://millercenter.org/the-presidency/presidential-speeches/april-23-1975-remarks-tulane-university

Johnson, P. (2013) *His and hers ... health care*. TEDWomen 2013, TED.

Miller, L. (2018) *Cultural Resilience*. Sydney MAD Mondays, YouTube. Available at: https://youtu.be/6LrTjycgc-Y

Nixon, R. (1978) *President Nixon's 1978 speech to the Oxford Union*. The Richard Nixon Foundation. Available at: https://youtu.be/1gl84ilI8Do

Reagan, R. (1988) *Address at Moscow State University*. Reagan Foundation. Available at: https://www.youtube.com/watch?v=1lutYGxMWeA

Robinson, K. (2006) *Do schools kill creativity?* TED2006, TED.

Sinek, S. (2009) *How great leaders inspire action*. TEDˣPuget Sound, TED.

Index